MY WORLD

THE LIFE AND TIMES OF A
CIVIL ENGINEER

With best wishes,

Peter Cecker

My World

The Life and Times of a Civil Engineer

by

Peter Ackers
MSc(Eng), FICE, FCGI, MASCE, MCIWEM

The Memoir Club

© Peter Ackers 2007

First published in 2007 by
The Memoir Club
Stanhope Old Hall
Stanhope
Weardale
County Durham

British Library Cataloguing in
Publication Data.
A catalogue record for this book
is available from the
British Library

ISBN: 978-1-84104-173-5

Typeset by TW Typesetting, Plymouth, Devon
Printed by Biddles Ltd, King's Lynn

Respice, Aspice, Prospice

The motto of Bootle Secondary (later Grammar) School

*To my dear wife, Margaret,
our children and grandsons*

Contents

List of Illustrations

Acknowledgements

Thanks are due to the editor of *The Independent* newspaper for permission to quote from the article by Clare Short, to the *Aeroplane* magazine for the illustration of the Bristol Freighter, to Solo Syndication for the Brabazon picture, and to the Office of Public Sector Information for two illustrations from the Severn Barrage Committee's report. I am indebted to Black and Veatch, successors to Binnie & Partners, for several photographs relating to projects. All reasonable efforts have been made to locate the owners of other copyright material. Acknowledgements will be given in reprints if the copyright owner comes forward.

My older son and my daughter also pointed out a number of errors and omissions, including one or two senile moments, but above all I thank my wife for her patience and support during the preparation these memoirs.

.

CHAPTER 1

The early years

I WAS BORN IN BOOTLE ON MERSEYSIDE on 30 June 1924 and to set this in context we have to go back to around 1880 when the Ackers family, a farming community from Halsall, near Ormskirk, Lancashire, had to respond to the rapid social and economic changes that were occurring. The convention had always been that first sons inherited whatever little wealth there was, and the Ackers family was fairly low in the order of things. For example, one from an earlier generation had nothing more than a weighing machine that was worth mentioning in his will. The Ackers were not land-owners. Also the farming land available to rent was very restricted so that descendants other than the first-born male could do no more than seek work as hired hands for other farmers or get apprenticed to some country trade like a blacksmith or wheel-wright – and even those jobs were limited by the amount of work available locally. But the Industrial Revolution was under way, export of manufactured goods and import of foodstuffs and timber were becoming important as the population grew and fewer babies died at birth. The Ackers family of necessity sought fresh pastures, and where better than the expanding port of Liverpool, the gateway to North America especially?

I am generation number twelve in the established genealogy going back to the late sixteenth century, the early generations being based in the Warrington area, but Halsall churchyard was the burial place of many of my forebears. Ackers does not sound a very English name so where did it come from? It is pretty uncommon here, and, until fairly recent years, most families lived in Lan-cashire. There are few Ackers in the telephone directory but there are pages of Ackers in the Amsterdam directory, for example. So are we of Dutch ancestry? We just don't know but it is almost certainly a Norse name. It is known that the Vikings coming down the west coast got only as far as Merseyside, but on the east coast they came much further south: the Danegeld. They also spread into

Denmark and Holland, and ultimately into France, becoming the Normans – who in turn invaded England! So the Lancashire line of Ackers could have come in several different directions, though there is now no evidence of the name in Normandy itself. We might therefore descend from the west coast Norse invasion. It is an interesting theory anyway.

It was my grandfather who moved to Merseyside, married and raised a family of five sons and two daughters, my father being son number four. The grandfather I never knew became a well-respected member of the community. He was a shipping clerk having joined that important trade of import and export. Bear in mind too, that this industry's expansion meant that dock facilities had to be expanded, and to provide the labour force with housing, Liverpool itself had to expand. Bootle, on the northern outskirts, was in essence a 'new town' with streets of red brick Victorian houses, mostly terraced, being built. With these houses went all the usual services, schools, churches, shops and a large fire station because of the great risk of fire in the docks and store yards for the timber trade. Grandfather John Ackers and his family found a pleasant house in Bank Road, not far from the railway which connected Liverpool to coastal towns and villages between there and Southport. The 1881 survey shows the property boom, as the adjacent street, Strand Road, was only half built. Just round the corner from Bank Road, 86 Strand Road was our family home for my first nine years, until 1933.

Grandfather had already died by the time I was born, and the double-fronted house was home to just my parents, my sister and me, and Granny, whom I only remember as a crotchety old lady in black bombazine, who occupied one of the two front rooms, overmantelled, horsehair sofa'ed where children were occasionally seen but not often heard! My father, whom I was named after, was a joiner. He had been apprenticed as a wheelwright, a trade which seems to have family connections. He had indeed been a wheelwright in the First World War, serving in the Royal Artillery, though, like so many soldiers of that war, he would never talk to the children about his experiences. Two of my uncles were in the timber trade, and another was also a joiner, so the whole family had fitted into the expanding role of Merseyside, most of the larger docks being in Bootle, rather than Liverpool. My father was employed by the Mersey Docks and Harbour Board, so was involved with large-scale dockside buildings, with their heavy timber roofs. Later he was a foreman, and during and after the war up to his retirement in 1953 he was in charge of the maintenance of the landing stage at the pierhead, a structure that is still (2007) in everyday use.

One feature of Strand Road was its tramcars. Being a relatively narrow road, it was a single track line with passing places, where the trams could clank past one coming in the opposite direction. Turn right from our house, under the railway bridge and you were in the shopping area, lots of modest shops selling anything you wanted: paraffin oils and Aunt Sally (the local household cleaner), tobacco, sweets and newspapers, groceries with butter impressed with a pattern on top, butchers, Chinese laundry, bakery where you could take your own cake or loaf to be baked, vegetables, shoes, stiff collars, cloth and towels, bicycles, all a stone's throw from home.

School, when I lived in Strand Road, was the local church school of St Mary's, no longer there, either school or church, because they were bombed during the war, as was so much of Bootle. I will come to that era later. All I remember of infant school was the first day, when we were given the task of pulling three-inch squares of material into separate threads to make stuffing for rag dolls. Mine was a very difficult material, taffeta perhaps, whilst others got very easy coarse tweed – most unfair! In those young days my sister Lydia, two years older, could escort me to school. Turn left, 100 yards down the street, past a tethered nanny goat, round to the left again for another 100 yards or so. Primary school was there too, where girls and boys went through separate entrances and kept strictly apart. It was there that my ability in maths was recognised and encouraged in a way that would not be countenanced today. I was sometimes put in with the class one or two years above to test me out. One of the standard tests in that primary school for the ten-year olds was a set of charts that could be rolled down in front of the blackboard, of 'tots'. These were columns of four three-figure numbers that we were expected to add up in our heads and give the answer while standing in a row at the back of the classroom – get it right and you moved up a space: get it wrong and you moved down! Of course things like multiplication tables had been learned early on, and there was stress on correct spelling and grammar. I was at St Mary's until I was nine when we moved house to a newly developed estate, to Vaux Crescent, just off Fernhill Road.

Boom times for Liverpool were well past; there was a lot of unemployment in the Depression years, and although at one time my father was on a three-day week, he was never out of work, and we were as well-off as most working-class families. No luxuries or expensive holidays, but always adequately fed and well cared for. The Strand Road house had become something of a burden by then, especially to my mother, Mary Grace Jones as was, showing the Welsh connection of so many Merseyside families. There were also many Irish people, but that was the time of strong religious prejudices so that the Catholic Irish did not mix much with the Protestant English and Nonconformist Welsh. My grandmother later moved to South Wales where she died, but to eke out finances the other front room, with the bedroom above, had been rented out to a couple, Bert Lynch being a fireman at the station down the road. We had a fire bell in the hall! When there was a callout, the bell rang, Fireman Lynch would rush downstairs if he had been off duty and sleeping, pull on his trousers, jump into his thigh boots kept behind the front door, grab his well-polished brass helmet, dash across the road, by which time you could hear the clang of the fire-engine as it left the station. If it turned our way, Bert would jump on as it slowed down and off they would go. All very exciting for a youngster.

Immediately opposite was Lamb's Timber Merchants, a very large store of timber which would be prepared and delivered to users by horse and cart. That was the standard transport system in the 1920s, and the heavy loads being stacked in the road just opposite would sometimes require three, or even four, shire-horses to shift them, horseshoes sparking on the stone setts as they tried to get the great load under way. We had a privet hedge, which the horses had their eye on. If one was left unattended, it would cross the road for a nibble. Once, when I was five, one came over while I was stood at the gate watching, and I was so shocked by this huge beast that I was unable to speak for several days, and my parents reckoned that it was this that gave me a bad stammer, which affected school-days, but largely cured itself as I grew in confidence.

Strand Road in recent years became infamous as the site of the abduction of a young child, James Bulger, who was later brutally murdered. The shopping area was so badly damaged by the bombing (though No 86 survived) that it was pedestrianised and roofed to become an early shopping mall, though rather down-market. The Bulger boy was taken by two older boys when his mother was in a shop, persuaded to follow them up Strand Road, turned right onto Stanley Road, past where the old Sun Hall picture house had been (happy memories of Saturday morning children's programmes of Laurel and Hardy and Tom Mix), along to Merton Road, turning left and so past the girls' secondary school where my wife, as Margaret McGeagh, had been a pupil, across the road to a reservoir, where the plan to murder the youngster must have been hatched, and on towards a disused railway line where the killing took place. The young culprits were ultimately convicted, have served their sentences and been released with new identities.

But in our day, Strand Road was a very pleasant area. The houses had tiny front gardens and backyards. Children were safe to go out on their own; you could go to Liverpool by tram or bus, or get the train to Waterloo or Blundelsands for a day on the beach, or even to Southport. During summer holidays, a penny ticket on the tram would enable you to ride towards Liverpool, change trams, and go out on one of the new radial boulevards where the trams were in a grassed central reserve to 'the end of steel' where you could walk into unspoilt countryside, sit at the roadside or by a field entrance to eat your picnic before the return trip. Life was very satisfactory for children even if your family was not well-off. There were some better-off Ackers nevertheless: one uncle was a bank manager with a detached house in Crosby, where the well-off preferred to live; one was a manager of a timber merchants, but another was a joiner like my father and one went to sea to exotic places.

So it was in 1933 that the family got the tenancy of a newly-built council house just off a major road further from the docks and industrial area. This was 4 Vaux Crescent, close to Fernhill Road. Our new estate of small three-bedroomed semis was one side and on the other side was a field of corn awaiting harvesting. The roads were of concrete, so an early taste of civil engineering was watching the navvies tamp the concrete with a heavy wooden beam with raised handles at each end. Within the next year or two those fields had in turn been developed for private housing, not much bigger than our council house, being sold for £550 or £600. We had neighbours of course, on one side there was Bill Cowey, who was my great friend as a boy, living with an aunt and uncle. Why was never explained to me and it didn't matter but I think he was the son of a girl 'in service', perhaps with her employer as father, because he got better Christmas presents than my family could afford! If I got a Number 1 set of Meccano, his would be a Number 5! We nevertheless enjoyed comparing our modelling skills when we weren't playing marbles in the gutter. Traffic then was so light that vehicles were the exception, and of course none of the council house tenants had cars.

Moving house meant moving school and my parents were keen that my sister and I should benefit from a grammar school education. Education was the means to escape from the low paid workforce to a real profession with much better conditions and an assured future. So passing the scholarship, later called the 11-plus. was the objective. It was this pressure that made my parents seek the best primary school for me, and so I went in turn to Hawthorn Road, Linacre Lane, Balliol Road and Bedford Road schools, finally getting to the one with the best record of scholarship success. So it was my last year at primary school that gave me the necessary entrée to grammar school, university and a profession, which meant considerable sacrifice by my parents in terms of buying the uniform and then supporting my absence at university.

Bootle Secondary School for Boys was about two miles from home and in those days all students walked to school, in rain, sun, fog or snow. I am eternally grateful for the excellent education I got there, with fond memories of school friends, many of whom went on to great achievements despite their humble start in life. I do not think there was much better schooling anywhere, and my thanks go particularly to Messrs Kay (physics), Smith (maths),

Gourley (applied maths) and the head, Dr Berbiers. I kept in touch with him up to his death: a dapper man who was well-respected in diplomatic circles as the Belgian consul. By the time I went to secondary school, I was exceedingly competitive. I wanted to be top – or nearly so – in all those subjects in which I had an interest, especially maths and physics. Religion and history were much less interesting – there was no logic in them! So I had my nose to the grindstone during my school years, but nevertheless was an avid reader of adventure stories, and took time off not in sport but in things like kite flying, because I could make my own, and even devised a carriage that would climb the string with outstretched wings, drop things like a bomber when it got to the top, fold its wings and come back down again. So the years from 1935 onwards had an established routine, while clouds were gathering over Europe and our lives were to change radically. My other hobby was collecting cigarette cards, and my father had organised all his work colleagues to pass theirs on to him. He was not a smoker, but he did like his pint of beer (or three or four) on a Saturday night with his pals at the Old Roan pub. During this period of happy childhood, my mother worked nearer to Liverpool, as a supervisor in a quilt-making factory, the Purax Company. Thus the family income was supplemented, but we were what would later be called 'latchkey children' – we did not have to wait till we were twenty-one to get the key of the door. This did us no harm but perhaps made us more self-sufficient.

The walk to and from school could take several routes in the largely gridiron pattern of streets but one route went past the gasworks. An amazing devil's kitchen of flames from the chimneys and through perforations in the brick walls, these spelling out *Ex fume dare lucem*: a useful example for our Latin teacher, which I believe translates as 'From smoke to get light'. There was a monorail system from the end of this retort house to take the coke out, and the little trucks had hoppers beneath that tipped the red-hot embers onto a long pile of coke. Also there was a blacksmith's shop en route where we lads could stop and watch them put the iron hoop on a new wooden cartwheel, getting the hoop hot in the furnace and slipping it over and then quenching it in a bath of cold water to shrink it on. In those days of heavy industry and coal fires, fogs were real pea-soupers, when you could hardly see the ground. The trams could get back to the depot with the conductor walking ahead with a torch, but all other traffic stopped. There were flares lit alongside the kerb to help pedestrians get home, and on one occasion I remember getting very disoriented when crossing the road close to the crossroads near the gasworks, and entirely missing the kerb on the other side, wandering a bit in the middle until I found my way again.

Holidays then were two weeks spent on my mother's cousin's farm, Trogog Uchaf, a couple of miles from the small cove of Amlwch, on Anglesey. We went there by train, changing to a local line across Anglesey through the place with the longest name in the UK, abbreviated to Llanfair.p.g. – but I can still say the full version I think! Dad only had a week's leave in the summer, but luggage in those days could be sent by rail in advance, door to door. So we urban dwellers saw the farming community at close quarters, mucky farmyard, cows to be milked, cream and butter prepared on the farm, rabbit shot for Sunday lunch, home-baked bread. Probably there was meat only once a week, though there was always a side of cured bacon to take a slice from. Home-grown vegetables would often be the sole food on the table, for example boiled potatoes and the soured milk left over from butter-making – not everyone's favourite. The peas were good! Oil lamps and candles provided the only lighting, and there was a two-seater privy set above a dry ditch at the back of the farmhouse. We could walk several miles to the creek at Amlwch to bathe, or from there get a bus to beaches several miles away with wide stretches of sand, picking blackberries from the hedge as we walked back to the farm.

One of our outings from the farm would be to visit my mother's Aunt Ellen Jane, who lived in the shop she once ran in the village of Pen-y-sarn. She spoke no English at all, but my mother still remembered enough of her Welsh to be able to communicate. My sister and I were enchanted by what had once been the only village store: the scales with its set of brass weights, the drawers with the names of contents on the front, e.g. sugar, flour, biscuits, butter etc., and a large weighing machine on the floor for the coal or potatoes. My great aunt had indeed also been the local coal merchant!

Bootle had a theatre as well as several cinemas, including the Metropole on Stanley Road. We would go to the pantomime as a Christmas treat, but quite recently I was reminded of another visit to this house of varieties by my sister-in-law, Margaret. She mentioned going there with her parents to a variety performance that included a naked lady! I suppose we were about fourteen or fifteen at the time and obviously our fathers had heard of something rather special, involving a display described by some euphemism like 'Classical Art'. This was an excuse for a bit of naked female flesh in artistic poses, one of which I remembered was called 'Gainsborough', when the model appeared wearing only a large picture hat, sat in an elaborate gilded frame. No movement of course, very discrete, and no pubic hair, probably with a piece of sticking plaster placed strategically too! Coming out of the theatre my mother's comment was 'When you've seen one, you've seen them all!' – but at least I had seen one in several very artistic poses, even if it was the theatrical equivalent of being air-brushed! So life as a child in Bootle had been very satisfactory for us but many were much less fortunate. It was the time of free boots for the poorer kids, and a free breakfast if they went to school before the rest of us, a bowl of porridge and a piece of bread and dripping. There were unemployed people around us, and what must they have thought of their lives. One lad just round the corner was nicknamed 'cloggy' because he had these free boots, actually clogs, and the only ones worn in our area were indeed the free ones. However, there were still children about in bare feet, especially in the summer months. In the winter, they could be seen with home-made carts, based on an orange box and a pair of pram wheels, going to get coke from the gasworks, the cheapest fuel available, especially if you collected it.

But in 1939, when I was fifteen, all our generation's lives were about to change. The September broadcast telling us we were at war with Germany immediately raised the spectre of air raids. The German contribution to the Spanish Civil War had been Guernica: would this happen to us? Of course we were in a strategic location, as Bootle contained the major Merseyside docks. So we were very quickly evacuated to Southport, some fifteen miles along the coast, a very pleasant seaside town. We all had our cardboard gas-mask cases around our shoulders, with very modest suitcases in our hands when we joined a special train at the station near to school, with a label pinned to our clothing saying where we were from. We arrived in Southport to be met by billeting officers whose job it was to allocate us kids to households who were to look after us for the duration. My first billet was not good as the rather poor family, willing though the lady had seemed, could not really manage an extra mouth to feed. My second billet was with a single man, of the Christadelphian sect, who were pacifists and conscientious objectors. Their humanity made them keen to help us evacuees however, and although my father had been a wheelwright in the First World War, he had no illusions about what war would entail and, I believe, fully understood the position of conscientious objectors and was grateful for their humanity. He never expressed any adverse views, and I was well looked after. It was Eddie Green who introduced me to mountains in fact. He had a motor bike and sidecar, and saved his petrol until he had enough to join friends in going to the Lake District one weekend when we climbed Helvellyn. The experienced ones then did Striding Edge but I sat at the summit. Striding Edge needed a bit more experience than I had of such terrain, though some years later I did it with my wife-to-be.

On my first arrival at his house, what I remember in particular was the smell of apples stored in the front parlour, so it must have been late autumn. He had a friend who was virtually blind – his eyes were glazed white – and he came from time to time pushing a small handcart with some vegetables he had grown. Despite his handicap he could tinkle a tune from the piano. Eddie Green was a pleasant host, and showed me one day how to catch a mouse. We were sat reading in the evening when he whispered 'there's a mouse over there'. There was a cast-iron fender around the open fire, marking the edge between the tiled hearth and the carpet. This fender was basically three sides of a rectangle, leaving about a three-inch hollow in the casting at floor level. He moved one end a few inches from the skirting and told me to watch. Sure enough, as the mouse worked its way around the room it spotted the hole under the fender and in it went. Eddie then pushed the fender back against the skirting. the mouse was trapped, but how to get it out?

He put his penknife-blade under one corner of the fender and explained that as the mouse ran round the corner, its tail would appear under the fender and all he had to do was to get it by the tail when this happened, as it certainly did. Grabbing its tail and lifting the fender a little he could take the poor creature out to dispose of it!

Later, I stayed with his aunt and uncle, with their son Frank Nettleton, also Christadelphians. It was a strange sect, being based on literal interpretation of the Bible, although the younger generation was prepared to accept some aspects of modern knowledge. Old man Nettleton was a die-hard, however, firmly believing the earth was flat, doing his best to convince me that all the evidence of curvature, such as ships disappearing over the horizon, was a trick of the eyes or bending of light in the atmosphere. He had a workshop in the back garden, and carved pipes for the Dunhill firm. A representative came from time to time to pick up the finished ones and deliver materials for another batch.

Strangely, however, evacuation proved to be a hit and miss affair, because the early war period became known as a phoney war – not much happened! Consequently the inconvenience of the school being evacuated meant that a decision was made to return to Bootle, and, although we were evacuated again to Southport later, for one reason or another we missed none of the severe bombing raids.

CHAPTER 2

Bootle's blitz

I<small>T WAS CLEAR EVEN BEFORE WAR</small> was declared that the Germans had a powerful air force and that they would have no hesitation about bombing towns and cities. Of course we were a long way from German airfields but the success of the German armies in France and the Low Countries meant that the threat got progressively closer. We had made what preparations we could, of course, with effective blackouts on the windows, and it was my job – being of artistic yet practical bent – to make a neat job of putting strips of sticky paper on all the windows. These were wooden sash windows with rather small panes, and the sticky tape certainly did its job when put to the test. Most important, however, was our air-raid shelter, an Anderson. This was of corrugated steel, delivered free to all houses with a garden, to be sunk half into the ground and covered with a mound of the excavated earth. It was Dad's job to dig the hole, where his little vegetable garden had been. This shelter was about 2 m (6 ft 6 ins) square, with an arched roof, and maximum headroom of about 6 feet. We were given bunks to 'sleep' the family of four, wooden frames with a rectangle of metal strips to lie on. We learned to rely on this when the raids started, but of course it was cold and damp in the best of circumstances. Other families without gardens were given Morrison shelters a bit later on. These were a replacement for the dining table, and were made of steel, the theory being that if the house fell around your ears you would not be crushed by the debris. Margaret's family also had an Anderson shelter, though I did not yet know her, as the girls' grammar school was about a mile from the boys' school.

Merseyside had its defensive array of barrage balloons and one was sited on the waste ground about 150 yards from the Vaux Crescent house, manned (or womanned) by WAAFs. There were also several permanent batteries of anti-aircraft guns, and some roving batteries that could be moved about. Later in the war, the sports stadium some half a mile away, where I used to go and watch cycle race meetings with my father, was taken over by a battery of anti-aircraft rocket launchers. There were about fifty of them, each being armed with two rockets at a time. These were about 6 inches in diameter and 6 feet or so long. The launchers were quite simple devices, mounted on a turntable and with provision to change their elevation. Presumably a central command gave them directions where to point, and a fuse setting, and then a signal would go to them all to fire simultaneously. They made a very characteristic very loud woosh, and the 100 explosions generated in the sky must have had a daunting effect on German pilots even if no planes were actually brought down. Little did we know when war broke out just what our town would be called upon to bear, and, strangely, the school children missed very little of it.

The best way to give an indication of Merseyside's ordeal is to quote from the official record of the bombing, *Front Line, 1940–1941*, HMSO, 1942:

> Liverpool was attacked 57 times and lost 520 of its citizens before its first heavy night raid on 28th November, when some 150 enemy planes attacked.
>
> Measured by number and weight of attacks and number of casualties, Merseyside must rank as Hitler's Target Number One outside London . . . The second was spread over two nights (three if some stray raiders are counted) a few days before Christmas. It was a heavy attack with many hundreds of deaths, widespread fires, and a good deal of civilian damage.
>
> Then, in the middle of March, came two nights of brutal raiding, bombs crashing upon houses and streets for eight hours the first night and six the second. The impact of the two raids combined, measured by the death rate they caused, was as great as the Coventry Attack, though the damage was less concentrated. With the shining water to help them the German bombers could not fail to drop many of their missiles on the docks themselves, and if they did great damage among densely packed houses away from, as well as close to, the waterfront, as they did in all the Merseyside raids, it was no doubt their policy of striking at the nerve and courage of civilians and disorganising their normal life to the utmost.
>
> The civil defence machine was by this time well tested and hardened and the services went about their work with full mastery, though the fire-fighters were hindered in places by the lack of water. In these earlier raids two boroughs had 30,000 of their houses damaged – about two in

every three. The people endured their ordeal with stubborn and
uncomplaining fortitude.

Six weeks later came the next, and for a long time the last, chapter:
Merseyside's 'May Week', the series of attacks on the docks and their
neighbourhoods that marked the first seven nights of May. Only two of
these raids were extremely heavy, but none was negligible. Between them
they killed 1,500 people. It was estimated that more than 2000 bombs fell
on land during the week and that the brigades fought over 1,500 fires.

The policemen and firemen who guarded the docks in May were
perhaps under as fierce an attack as any men in the whole course of the
onslaught on Britain. High explosives and incendiaries fell in great weight
upon the dock basins, the quays, the ships moored at their sides, and the
store sheds hard by. There was damage, but the marvel is that it had so
light an effect, and for so short a time, on the working of the port.

Policemen moved arms and ammunition to safety from blazing sheds
with their own hands or put their shoulders to trucks laden with shells and
forced them away from spreading flames. Firemen fought all night to check
the fires on a blazing ammunition ship. All these men, every instant for
hours on end, were consciously staking their lives on the race they were
running against time, the threat from flames and falling embers, great as it
was, being less than the chance of immediate explosion at their sides if the
flames moved too fast for them. Volunteers from among the dockers
worked to unload special cargoes while the bombs fell, and cleared in
record time some naval vessels which the admiralty wanted to move from
the danger area.

At the end of this week of desperate risk and heavy labour the docks
were working: handling a reduced volume of traffic for a time, but
working. As blocked roads and cratered railway tracks were reopened, and
the chaotic reminders of fire and bomb cleared from the quaysides, the port
moved back towards normal working. The enemy had done his worst to
Merseyside for a week, and there was much to do outside the docks as well
as inside. Vehicles, civil and military, were mobilised from the nearer parts
of Lancashire, and for days the work of clearance and repair went quickly
forward. In all the Merseyside raids a total of over 150,000 houses were
damaged.

Forty thousand homeless people were billeted inside the city in one
week: others moved to Rest Centres and billets on the outskirts and in
nearby villages and towns. It might have been feared that sections of the
dockside population would show some weakness of nerve. But every
testimony agrees that there was no sign of it: they moved to their
temporary outstations cheerfully and in cold blood, grateful no doubt for
the prospect of some nights' peace and quiet. The police, who have had
past reason to take a cool view of some of the dockside neighbourhoods,
spoke in praise of the way they had stood up to their ordeal.

Some of the housing districts were affected worse than others. In one section almost every warden was homeless after the first few nights. They took their turn of sleep in shelters or Rest Centres and worked straight on, day after day, part-timers as well as whole-timers. Other services had still worse ordeals to face. Of the First Aid Party Depots, only one was left unaffected at the end of the raid. A bomb fell directly upon an Ambulance Station, killing 17 drivers at a blow. One of the divisional Control Centres had a bomb through the middle of its ceiling. Happily it did little injury: eyewitnesses said it 'tore its side out on a girder and went off like a squib'.

Some thousands of houses in Bootle were roughly handled by blast and bomb splinter, and parts of the borough looked very untidy towards the end of the raids. But the inhabitants were not to be driven from taking things too seriously. On the morning after the last raid – no one of course then knew it was the last – an observer, picking his way through the streets, saw women at work in the habitable houses, and in a good number that did not deserve that description. They were scrubbing the steps, polishing the door handles and cleaning the remaining panes of glass, as they had done before the raids started and are no doubt doing to this day. Another observer who knows Merseyside well, summed it up for an enquirer a little time after the raids. 'Of course there's no doubt,' said this authority 'that if Jerry kept up continuous raids night after night on place like Liverpool a lot of people would disappear' 'And when would they come back?' The authority replied 'Next morning.'

This direct extract from a government document published in 1942 is a very good description of the severity of air raids, but it does not record the personal involvement and feelings in all this, and of course the government was rightly cautious about giving too much away about the damage caused and locations of it. The ammunition ship continued to explode for perhaps twenty-four hours with flames leaping up and the sound of explosions heard all over Bootle. It was indeed in Bootle, in the Huskisson dock which was so completely ruined that it had to be filled in. In fact the major Merseyside docks were – and still are – in Bootle so the borough was particularly hard hit. We were lucky to live some two miles from the docks, so we were not in the worst hit parts of the town and lived through it all, though our families had much to contend with.

My father, as a foreman joiner on the docks, had much involvement of course with getting the dockside buildings cleared enough for the port to work again. Despite a sleepless night, he would set off early to work with the prospect of having to walk much of the five miles or so involved, anticipating that roads and railways would be blocked and the overhead tram cables down and switched off. The overhead railway that ran alongside the docks would almost certainly be out of action. My mother, too, would have similar problems getting to work, now making kapok-filled Mae West life jackets rather than eiderdowns. My sister, Lydia, had left school just before the war to go to secretarial college and worked in a bank in the city. One of her memories is arriving at work one morning to find the bank was no longer there, just a hole in the ground with bank notes wafting about and in the water-filled crater where the vault had been. Her job was to sieve out the wet notes as part of the salvage operation! Later she was in the Women's Army (WRAC) with a khaki uniform and seconded as secretary to a general in Chester where much of the planning for the Normandy invasion was carried out.

I had to keep up with school-work and homework because it was in the autumn of 1940 that I went into the sixth form, and who knew what might then follow. So it was a case of picking my way through streets, some closed because of unexploded bombs, most bomb damaged, fallen overhead tram wires to step over and, of course, everyone else doing similar things to try to get on with their lives. One day, crossing the waste ground near home on my way to school I joined others looking over the railway fence to see an unexploded land mine, with its green parachute draped to one side. These were large bins of explosives, dropped on a parachute to stop the weak case from fracturing on impact, the purpose of which was to explode at ground level to cause very widespread damage. Lying, shivering, praying in our shelter we learned the different sounds of a bombing raid. The crack of the anti-aircraft guns, the whistle of falling shrapnel, the crunch of bombs and, perhaps most frightening of all, the flapping of the parachute on a falling landmine.

Our own house survived, though we lost slates from the roof through falling shrapnel and the main living-room windows were blown in as a whole unit by blast on three occasions. Not one pane broke, so that sticky tape did its job on those small panes. With some help my father was able to put them back in place so the house remained inhabitable. Margaret and her family were not so lucky. It was the raid just before Christmas 1940 that their terraced house was severely damaged with most of the roof gone, windows broken and ceilings damaged. The last straw however was an unexploded bomb nearby so the warden made them evacuate the house within five minutes or so, carrying what they could. They were then found a newer house further from the dock area in Litherland, and it was there she lived when I first got to know her.

For security reasons, the radio news never once mentioned Merseyside or Liverpool as a place that had been bombed. Even though Jerry knew perfectly well where the bombs had dropped, it was never more specific than 'a place in the north-west' and the result of this was that even now few people with wartime memories realise that Merseyside, with the Bootle docklands, was the next most severely attacked place in the country to London. People remember Coventry, where 1,236 were killed, though that was about the number killed in Bootle with the Merseyside total being about 4,500 deaths. There was devastation in the Rimrose Road, Derby Road area in particular (Fig. 1). Three-quarters of all houses in the borough were seriously damaged at least once. On 28 November 1940, 150 bombers attacked, 500 on the nights of 20, 21 and 22 December, and 800 in our 'May Week', and these were not the only raids. Small wonder then that none of us felt much sympathy when the Germans themselves suffered 1,000 bomber raids much later in the war. Obviously there were many instances of bravery during these raids, and one of the lower sixth boys, Ron Hayes, was awarded the BEM for his role during the height of the attacks as a messenger boy, riding his bike between the various emergency services to deliver messages when the phones were out of action.

By May 1941, I had just been made a prefect at school and as such could use a small room on the first floor of the central tower. I borrowed a key and got a duplicate cut at a shop on Stanley Road for a precious half-crown. But I never used it. Next day, arriving at school after a very bad night, there was no tower and much of the school was still smouldering, having been hit by incendiary bombs during the raid. We would be evacuated back to Southport later that day. I had to walk to my mother's workplace, several miles away, to tell her, as there were no trams or buses running –

and it was well before mobile phones existed of course. Nor were the public phones working. Back to collect together the essentials for leaving home again, and back to the station for a special train. Thus much of my time in the sixth form was spent in Southport.

CHAPTER 3

The sixth form and the 17-Plus Club

A S MY SCHOOL WAS FROM A working-class area, most of the students were expected to leave at sixteen with a pretty good basic education so that they could expect to get a job leading to a career in business, in banking or in an office in one of the many businesses on Merseyside. It was a relatively small proportion that were considered to gain much benefit from taking their academic training further. There was always night school for those who left at sixteen but wanted to get training for some special field. Only six of perhaps 100 in my year stayed on, and there was a similar number from the year before and after to give a total of about a dozen between upper and lower sixth forms. This small band of six became especially close friends when they were all away from home and families. Our school in the old chapel in Part Street, Southport, had a stage in the main hall with black curtains, and this was allocated for use as a sixth form room, and provided with several tables to work at. We were a mix of arts people and those into science – including me because maths and physics had been my special subjects. We had to work a lot on our own, using the text books recommended, because our teachers necessarily spent most of their time with the main classes. My subjects were maths, applied maths and physics, though later I did some intensive work in chemistry. We were all working for A level school certificates, having got high grades at O level in our chosen subjects, and, though we didn't really know quite what career we might choose, the presumption was that there was a chance of university if we managed to get one of the awards that the local borough council had in its gift, providing maintenance grants. Bootle Borough Council was very good and far-sighted in this regard. If one's grades were good enough, the universities would give free tuition in those days. I doubt if any in my year could have gone on to university without the benefit of not paying fees, and some assistance with other costs to eke out what one's parents might be

19

able to provide. There were features about that sixth-form evacuation which helped shape our whole lives. This was a group of clever youngsters with quite strong characters. We also had a few girls in our boys' school: just a small number had declined to join their colleagues in Herefordshire and were allowed to join the boys in Southport.

One particular friend was Jack Morgan, who was later to be our best man. He specialised in languages, French being his number one choice, with our Mr Porter as his tutor. He went on to university and later did a Japanese course, a language that would be important because Japan had joined with the Germans in the war. Sadly, we lost touch after the war though I think he went into the antiques business. He was one of the few who smoked (not in school of course) and was a bit of a lady's man! My other very good friend was Arnol (no 'd') Banks, also a maths and physics student. He was an unforgettable character, with remarkable abilities in many fields and some extraordinary interests. For example, back in the Bootle days, we would go together to the public library where I would peruse Jane's *All the World's Aircraft* or its twin, *All the World's Battleships*, while he would get special access to the locked bookcase with the *Egyptian 'Book of the Dead'* – though he then knew all about embalming and mummifying, I don't think he had undertaking in mind as a career! I think of him now as a polymath, he knew a lot about many things. If the electric motor in a vacuum cleaner failed, he could rewind it. He could brew some sort of alcohol, as I found out when standing next to him in a line on the sports field. There was a stoneware bottle in his back pocket with a string tying the cork in but it was nevertheless foaming due to the fermentation; (he and I were alike in hating anything to do with sports or PT). He was into explosives, making black powder. This he ground in the fireplace in their basement so that if there was an explosion it would be contained! He lived in a terraced house and had to go down an entry to the back door when he got home from school. A neighbour had a bad-tempered dog that awaited the sound of his hobnailed boots and then came out to worry him, ultimately biting him. But he made home-made guns using a tube reinforced by being cast into a block of lead, so one day, when asked whether he had got past the dog the previous day, he replied, 'It won't trouble me again. I shot it last night!' He made rockets using glued-together sheets of paper as the tube. This got him into trouble in Southport because the police or coastguards thought someone was signalling to a German submarine and found a discharged rocket case. This was made of sheets of paper saying 'I must not be rude to prefects' again and again – fifty lines he had given to someone from the lower school, hence it got traced and the science master knew who the culprit was likely to be. One day, with the sixth form on the stage behind the curtain before morning assembly he brought out his latest toy, a model cannon. He demonstrated how to load it, ram down the powder, pop in its lead bullet so it was all ready to fire – and that is what he promptly did when we all said, no, not here, not now. A huge bang and the bullet went through one of the wooden screens separating a classroom from the hall and out through a window into the street, filling the stage and hall with acrid fumes just before assembly. We scarpered, of course. He was in due course to join the police and became an inspector! I also remember him as having an interest in space. Arthur C Clarke had set up the British Interplanetary Society, and at the age of seventeen, Arnol was member number thirty-six. He brought in the cyclo-styled proceedings so we knew then about the requirements for getting into orbit, the possibility of communication satellites, the size of solid fuel rocket that would be needed to reach escape velocity, and how artificial gravity could be achieved by spinning a doughnut-shaped space station. He was such a remarkable youth that the memory of his range of interests and his exploits could never be forgotten.

Then there was Joe Brenner, a Jew, but there was no concept of prejudice in our school. Your religion mattered only in that non-Christians were excused from prayers at morning assembly. His father was a tailor and made several suits for my father. He was good at most subjects so was a class rival in a sense. He went in for medicine and the only post-school news I heard of him was that he was running a clinic in New York for poor people (though I can't remember the source of this).

Another good friend was Tony Herbin. He was primarily a geographer but he came from a musical family. The tale was that his parents had performed at the Scala Opera House in Milan, but in fact their professional music was at a more mundane level, his mother being English and his father Italian. Probably they performed at one of the Liverpool theatres, as conductor and *chanteuse*. He had access to a fine collection of records (78s in those days) and when in Southport we had a weekly music club where he would play some to us. That is how my great interest in classical music was born. One memory of youthful pranks is what happened when he played Ravel's 'Bolero'. We did a conga-like dance round the stage in rhythm and I was very worried because the stage floor seemed to bounce to the rhythm and gave up clouds of dust that I though might presage its collapse! Tony became a lecturer at an African university, and actually married the sister of someone in the lower sixth. In effect he had to retire through ill health and returned to the UK to live in an apartment in his brother-in-law's house in Bootle where he died some years ago, I think from diabetes.

A very good sixth-form friend was Joe Hughes (his Christian name was Robert and he had a brother Trevor in the lower sixth). He was also a science and maths student, and after university worked for Shell in the Wirral as a tribologist, the study of lubrication. Sadly, his life went awry somehow and he committed suicide when he was still quite young. All of this group with the exception of Arnol Banks went on to university, getting some sort of award from the Corporation (Fig. 2).

As the age of eighteen approached and attention turned to a future career, I had thought of civil engineering. My father's work for the Mersey Docks and Harbour Board, now as a foreman, was essentially dock engineering and structures. There were heavy timber roofs to the dock sheds he was involved in building and this was almost civil engineering, and he was familiar with the work of the Drawing Office and the Chief Engineer's Department and what they did. Careers advice at school was on the whole left to one's teachers and my applied maths teacher, whose name was Bill Gourley, had a brother or cousin who was a partner in the consulting firm, Binnie, Deacon & Gourley, specialists in water engineering. I presume it was through help from him that he advised me to try to get to City & Guilds College of London University, part of Imperial College. Also there was a remote chance of getting a scholarship that would cover the fees and provide a grant. This was the Royal Scholarship, only two of which were granted each year on the basis of a competitive exam in maths, applied maths, physics, and chemistry as a subsidiary. I was not doing chemistry in the sixth form so I was involved in especially intense mugging up of chemistry over just one term to have any chance. Anyway, performance was good enough and the news came back in due course that I had passed, supported by a score of 100% in maths, which, after all, was my strongest subject. The local council came up trumps too with a local scholarship providing an extra allowance for maintenance. I was on my way and going to London to continue my studies!

The boys in the lower sixth did just as well a year later in getting the necessary grades to be able to go to university, and two or three became professors or senior lecturers or in senior positions in their careers in the course of time. Jim Scott was one, and he finished up as a professor of geography at Washington State University in the USA. We met again when he returned to the UK for his eightieth birthday, and also met again his contemporary Ken Young who was an engineer and, to my great surprise, he had worked as a senior works manager alongside my university friend, Frank Sibley, in the north-east. I had been best man at his wedding. It is just one of many instances demonstrating that this is not such a big world after all! Gerry Galletly became professor of mechanical engineering at Manchester and I met him when we were both in our fifties when I gave a lecture there. Pete Filchie was a civil engineer who lectured at a Liverpool college. In fact the friendship between upper and lower sixth extended much beyond school hours, because being on our own in Southport meant that many after-school activities were organised, in effect, by our own youth club. When back in Bootle during holidays for example we mostly belonged to the 17-Plus Club, an offshoot of the national 18-Plus Clubs. That was open to the girls as well, of course, and many boy/girl relationships grew up apart from me and Margaret. There were many non-sixth formers and non-grammar school members, with sisters and brothers being introduced. In fact Ernie, my brother-in-law met his wife Margaret there. Leading lights were the Hardacre brothers from my school. We organised

youth hostelling holidays in North Wales, the Peak District and the
Lake District, getting much exercise fell walking and mountain
climbing (Fig. 3). There were weekly meetings, many with outside
speakers including Bessie Braddock, a rampant Liverpool socialist
councillor and even Malcolm Sargeant who was then conductor of
the Liverpool Phil. who spoke about Sibelius' 'Swan of Tuonela'.
We had premises in an abandoned cricket pavilion in one of
Bootle's recreation grounds, which we had somewhat restored and
redecorated with gifts of paint from someone's father in the trade.
We played table tennis, or cards or Monopoly, very popular at the
time. We organised it all ourselves and if short of funds would run
a rummage sale in a church hall somewhere. There was a great
atmosphere arising from the extended friendships formed and this
self-sufficiency augured well for our futures in the wide world.
Margaret and I continued our membership of 17-Plus until we
were married when there were other priorities arising from work
and then moving out of the area.

It is surprising that although selection at age eleven was dropped
some thirty or forty years ago, doing away with the grammar
schools in the borough, there is still an Old Bootleians organisation
which publishes an annual magazine and keeps us in touch with
the now far-flung school colleagues. This is a remarkable achievement, and indicates how much we valued our education and the
friendships forged then.

CHAPTER 4

To London and university

I HAD NEVER BEEN TO LONDON BEFORE and so this was a step into the unknown. It had been possible to arrange digs near the college, in Eccleston Place which was about half a mile from college. Arriving at Euston by train, I was faced with finding my way to South Kensington with my suitcase, with no prior knowledge of the underground system and we working-class folks had not been brought up to use taxis which were considered an expensive luxury. Anyway I found someone who volunteered the information that I had to find my way to the Piccadilly Line – or was it the Circle Line I took. Too long ago to remember but my small room was on the top floor, really in the attic, with a little gas radiator with a lid at the top that opened to show a tiny gas ring, just right for making a supper drink. Breakfasts and evening meals were provided and we students had a table to ourselves in this boarding house. There was food rationing of course so our books of coupons were handed over to cover the meagre allowance of butter, sugar, cheese and meat.

I actually started at Imperial College a month before term proper opened because to be an engineer I needed a knowledge of machine drawing. The few who came via apprenticeship schemes and perhaps a Whitworth Scholarship had already done this but the rest of us needed to know all about projections and screw threads and the like, hence this month extra. I was good at drawing anyway so this was no problem to me. The first day of real term, we had someone from the armed services who explained our role in no uncertain terms. We had been selected to be future scientists and technicians and were therefore exempt from the call-up, but, as these specialists would be so important to the war effort we had no choice in the matter. Anyone who felt like volunteering despite their exemption would be brought back to finish the course. The degree course would be squeezed into two years so we were going to have to work long hours and hard and any thought of university being an easy few years was out of the question. So I was at university doing my civil engineering for only two years.

The first year covered mechanical and electrical engineering as well, because our later careers might have required those extra disciplines. We had lectures every weekday, interspersed with practical work in the workshops or laboratories, and study periods when we worked under supervision doing exercises or drawing up schemes. So we covered such things as practical concrete mixing, testing steel samples until they broke under tension in the powerful machines, hydraulic experiments on weirs in the water lab., surveying in Kensington Gardens, the efficiency of internal combustion engines and tests on electric motors. Thus it was quite a wide and very practical training, with some excellent professors and lecturers teaching us all the theory, including more advanced maths than we had learned at school. We had used slide rules in the sixth form and these were essential in doing structural designs etc. Our maths lecturer was so nearly blind that he would write long equations over the width of two blackboards on vertical runners and didn't know that we were utterly foxed when he pushed one side up to make more room!

Of course many of my student colleagues were from a different part of society than I came from, many from public schools. Class differences were never a barrier and, being the one with the highest entry qualification in the group, I was always treated with respect. They would ask me questions in the study classes rather than the tutor: was this to hear my strange 'Scouse' accent or because I could explain things clearly? I wonder! Many of them came from wealthy and reputable families. Stopes-Roe, for example, was the son of Marie Stopes, the outspoken feminist of her day who pressed for better information about birth control. The Roe in his name was from A V Roe, the aircraft manufacturer, Avro. One student had even been able to build his own wind tunnel in the conservatory of the house where he lived to test model aeroplanes! Most of the students were fee-paying as their parents were of ample means but all had to have good A level results to gain entry to probably the best technical university in the country (Cambridge was more for pure sciences and arts, and Oxford's reputation was in the arts too). There were quite a few grammar school students with their local awards to support them and my best friend there was Frank Sibley, who was from Manchester Grammar School, and was later to ask me to be his best man. Peter Andrews was another good friend who was from South Wales and came with a Whitworth Scholarship.

We all had extra duties for the war effort. Many of those from public schools had been in the Army Cadets or the Air Cadets so they were seconded to serve in the Home Guard or continue their air force training. I was told to become an air-raid warden, reporting to a unit in the basement of the block of mansions opposite the back of the Albert Hall. I was issued with a black battle dress uniform, a black tin hat with a white 'W' on it, and a greatcoat. It was there that I had my closest shave of the war, because although Merseyside's air raids were virtually over, London remained very vulnerable. There was a battery of heavy anti-aircraft guns in Hyde Park, and on duty just by the Albert Hall one night there was the whine of falling shells just after the anti-aircraft guns let off a salvo. One, two, three, four, five getting ever closer. I ducked by the wall not far from a telephone box. One of those shells actually knocked a pinnacle off a corner of the Albert Memorial, while the next hit the road, just beyond the heavy granite kerbstones and exploded. This probably saved me from injury, because it deflected the shrapnel upwards, breaking the upper panes in the telephone box and pock-marking the wall above me. I was shaking like a leaf when I went to report back to base to the head warden, an ex-colonel.

During the summer vacation between our two years we were required to get practical experience whilst helping the war effort. I was allocated to work on an opencast coal mine near Wakefield, where I had digs in a cottage in nearby woods, so it was very much an outdoor period. The opencast workings consisted of taking out the soil and soft rock above the coal seams using big scrapers pulled by caterpillar tractors, which dug up perhaps thirty or forty tons each go, transporting this material to a spoil heap. Then drag-line machines would dig up the coal seam, loading the coal into lorries that went off to power stations or wherever. My job was to assist a Welshman in surveying because the contractor was being paid so much for every cubic yard of spoil and every cubic yard of coal he excavated. Accuracy of surveying was all important and then it all had to be plotted on plans and the volumes worked out, using an instrument called a planimeter to measure the irregular areas involved. This was all quite enjoyable when the weather was OK and I learned a lot from the practical engineer I was with.

I was still doing my duties at the Albert Hall warden's post in the second year at college, but the South Kensington digs were not available so I found some in south-west London, in West Norwood, with a train journey to get to college each day. This was in a pleasant semi, the home of an elderly lady with two spare rooms. The other was occupied by Squadron Leader Boyce, a New Zealander on duty at the RAF headquarters in Whitehall. I remember being surprised at the spring beauty of the almond and cherry blossom on the trees flanking the roads: we had not seen much of that in Bootle! My good friend Frank Sibley was in digs not far away and we always met on Saturdays, usually finishing up in the cinema, if there was anything worthwhile on.

If there were air raids when I was at the digs, I considered myself to be available for duty, and there was a phase of fire-bomb attacks. Like most students, however, my philosophy was to stay in bed to sleep if I could, expecting to just have time to slip out of bed and lie alongside if I heard a bomb dropping close. During one fire-bomb raid I heard the sound of falling incendiaries so put on my greatcoat and tin hat to see if any were close enough to put us at risk. By then the Germans had made a proportion of their fire-bombs also anti-personnel devices, and the information was that they should not be tackled for seven minutes after impact, judged to be the maximum delay on the fuses. Seven minutes being up I left the house. The only incendiary in sight was in the road so could be left to burn itself out, but I could see by the glow that there were others in the next street around the corner. There was an auxiliary fire station there and the volunteer crews were stood outside wondering what to do when I was met by a man saying there was a bomb on the roof of their block of flats. Could I help? All sizeable buildings had a rota of fire-watchers, and theirs was getting organised when I got to the top floor of either a four- or five-storey block. Someone had got a ladder up to the roof access hatch so up I went. I was given the stirrup pump as they thought, wearing a warden's helmet and being young, I was best trained to deal with the problem. All households and buildings had their hand-operated stirrup pumps which operated from a bucket of water. You needed a team of three or more to use them. One to direct the jet, one to pump and at least one but really a team to form a chain of water carriers. The procedure for the hose-man – me – was to take aim when lying down but to hide behind a bin lid held in the other hand when actually directing the spray onto the bomb, in case it burst in a spray of burning phosphorus. The pump-man was as far away as allowed by the length of hose. We were working efficiently as a team when the message came that the auxiliary fire crews had got one of their trailer pumps working and were about to send a jet up to the roof. We should leave them to it, which we were very happy to do. This was just the spirit of the day. Urgent decisions were taken without delay or discussion. People who had never met before and would never meet again probably just did what was necessary and some sort of leader would quickly emerge.

We were now about the middle of June 1944. There had been rumours of German secret weapons for some time, but most people, like me, took them as propaganda. I think it was a Friday night about 13 June that we learned that it was not just propaganda. There was a raid during the night and the planes we heard had a different sound from the usual one-second throb of the normal twin-engined German bomber. The explosions they produced were earth shaking too. I met my friend Frank next day, enquired how he had got on during the night – a noisy night but OK. We met up with a college friend just by chance and he was still shaking and told us the news: this was Hitler's secret weapon, which came to be known by its German code as the V1. He heard the thunderous crackle of an aircraft engine, sat up in bed, saw the flames from its exhaust as it passed his window, pulled the eiderdown over his face and bang, the windows shattered, the wardrobe was stuck with splinters of glass, he cut his feet getting out of bed. The flying bomb, the forerunner of the cruise missile, had flown into the pair of houses forming the end of the cul-de-sac, demolishing them and the neighbouring houses and killing everyone living there. No wonder he was shaking; and that

was the end of my scepticism about the German ability to develop
a completely new weapon. Being to the south-east of London, we
and all the other suburbs there were very vulnerable as many of the
V1s fell short of central London. They had a timer which switched
off the engine when their speed should have got them somewhere
over a populated area, when they dived into the ground to explode
on impact giving a great deal of blast damage.

Time for our finals was approaching, however, which were
taken in the Imperial Institute Buildings at Imperial College, only
the tower of which remains. Instructions included that if the
air-raid sirens sounded, we were to keep going. There was no
question of us going to shelters, but if we heard a bomb coming
close we could duck! If there was a direct hit, too bad! Luckily
there was a lull that week. Of course the V1s were not restricted
to night raids and when you heard one, you listened for the engine
to stop and if it passed you by it was some other poor soul's turn.
I well remember being on a bus going along Hyde Park Road
when we heard a V1. The bus conductor leaned out at the back
and watched it fly past and then shouted that we were OK. You
can't imagine that attitude being allowed today with the police
seeming to stop ordinary life and traffic and keep it stopped to
avoid the slightest risk.

Before the course came to the end, the Germans had unleashed
their much more advanced secret weapon, the V2, a ballistic missile
based on liquid-fuelled rockets with sufficient range to reach
London. There was no defence, no possibility of warnings, just a
double bang which, if you heard, you knew you were safe. Some
misinformation was put out to persuade the enemy that they were
all overshooting the centre of the capital and this may have caused
them to slightly shorten the trajectory to compensate. Anyway, it
was the east and south-east suburbs which really suffered. By then,
the V1 attacks had ceased. They could by shot down by heavy
concentrations of anti-aircraft guns and the most modern Spitfires
could just about catch them and tip them over to crash early. But
the V2s were only defeated when their launch sites or supply route
from the underground slave labour camps where they were made
were damaged by Allied air raids. When my course was finally
over, I could take a very short holiday at home before coming

under the scheme for the direction of labour, which affected every adult. I had left a small suitcase with text books in it with Squadron Leader Boyce to take care of with the intention of collecting it when next in the London area, but when I went for it a couple of weeks later, getting off the train at West Norwood the landscape was quite different. There was a wide open space, at the very edge of which was my old digs. Windows had gone, the front door was propped against its frame with a note on it. 'Would Peter Ackers please go to the RAF headquarters in Whitehall and ask for Mr Boyce.' The lady who had lived there had also fortunately escaped with her life. That was a V2, so I was still lucky.

Having started with a very good education in science and maths from school, I had no difficulty in keeping up with the work at university and I had learned good examination technique early on so finished with a good degree and was awarded the Henrici Medal for mathematics, the Unwin Medal for overall performance, to add to those awarded after the first year, the Coopers Hill Medal and John Samuel prize; (later I received medals from the Institution of Water Engineers and the Institution of Municipal Engineers for technical papers.) These medals include some fine examples of Art Deco design as well as traditional, perhaps Victorian designs (Fig. 4). Of course, wartime meant that there were no opportunities for post-graduate study or university research, so my whole career was based on just two years actually at university.

CHAPTER 5

What did you do in the War? Research and design for the aircraft industry

I WAS TOLD TO REPORT TO THE National Physical Laboratory in south-west London and digs were found at 1 Bridgeman Road, where several recent graduates were also staying as well as the older residents. Actually the laboratory was on the other side of Bushey Park so a pleasant walk across the park was needed to get to work. NPL had many departments covering different sciences and technology, and, as the premier government research establishment at the time, their attention had turned to the requirements of war. Radar was being developed there but as I was an engineer, my job was to work on the testing of materials at high temperature. It was understood that this had something to do with aircraft engines but it wasn't until later that I realised that it was largely to do with the special steels that would have to be used for the blades of gas turbines – the Whittle engine being the first British one. The steel was nimonic 80, a nickel alloy. The samples were in an electric oven to heat them to a pre-set temperature, whilst they rotated on a horizontal axis, driven by an electric motor. Thermocouples measured the temperature, the voltage they generated being taken out via slip rings. All the while they were being bent by having a load attached to the outer end through a ball bearing. Cleaning the slip rings and oiling the bearings was important whilst waiting for them to break as the fatigue due to repeated reversal of stress made them increasingly brittle. Some aluminium samples were also being tested in similar fashion but not heated. Of course it was later an incomplete understanding of – or perhaps failure to apply – this sort of research that lead to the crash of the Comet airliners.

But this was the work of a Junior Experimental Officer whereas I was there as a Junior Scientific Officer. I felt my talents were not being used properly and complained to my boss. He found me another more challenging task, how to design struts which were made of a sandwich of two sheets of steel with a lightweight plastic core between them. There had been some research on the methods of possible failure under load and my job was to find a way to bring it together into a method of design. This led to my first published paper written when I was still twenty. But I was only at NPL nine months before being transferred to the Bristol Aircraft Corporation, in Bristol as you might expect. It was in late 1944 I think that I observed an atmospheric phenomenon whilst walking early one morning to work across the park. It is so rare that it might only be seen once or twice a century, and is what around 1915 got the name the 'Angel of Mons' because it was seen over the battlefield. It is due to particles of ice high up, intersecting the sun's rays and generating mock suns. One or two mock suns occur fairly commonly in certain weather conditions if you know where to look (though surprisingly few people actually see them, so many are relatively unobservant). But the full development generated mock suns above as well as on either side of the real sun, which then is almost invisible. Each mock sun has a halo, which intersect as the vertical and horizontal arms of a cross, centred on the true position of the sun – and of the cruciform pattern – that led it to be called the Angel of Mons. There were over 500 bombers with their fighter escorts very high in the sky further to the east, probably American though I think British bombers had been out overnight too. There was obviously a serious attack on the German positions in France as the advance into Normandy from the established beachheads was progressing.

The head of the department at NPL was a Dr Wittrick, later Professor of Mechanical Engineering, whom I happened to sit next to at a lunch at Birmingham University. Surprisingly he remembered my name after thirty or so years saying he wondered what had happened to me. I was able to explain that after the war I took up the civil engineering I had been trained for.

I therefore went to Bristol as I was directed, and reported to the BAC (Bristol Aircraft Corporation) at the design office which was actually in the zoo! Lodgings were sorted out at Westbury-on-Trym, with an elderly widow who also had a young woman lodger. People of working age were generally away from home then, and in effect people with spare accommodation had to make it available to migrant workers or school children. This was an out-of-city old village and these digs were in an old cottage. Weekends provided the opportunity to walk in the woods at Blaise Castle or to meet work colleagues in Bristol, for a drink perhaps at the ancient Landogger Trow pub, or a game of snooker, or just to explore the large open area, the Downs, or the gorge with the wonderful Brunel suspension bridge. I was a junior stress-man, involved in the actual design of the aircraft then under development in terms of working alongside an experienced design draughtsman, who started the process off by looking at what was needed in one section of the plane at a time, within the general framework and outline shape supplied by the top designers to meet some requirement of the Air Ministry. The first one I worked on was the Bristol Buckmaster, a twin-engined fighter-bomber, and it was the control system then on the drawing board, a system of wires from the cockpit back to the rudder, elevators and ailerons on the wings, the wires turning corners via levers.

I was in at a much earlier stage with the Bristol Freighter, a short take-off and landing heavyweight carrier really being designed to service the army when it got to the stage of expelling the Japanese from south-east Asia, where it would have to operate from rough jungle clearings, a more efficient replacement for the American DC2, and able to carry a light tank or Bren gun carrier, or perhaps up to fifty fully armed soldiers. My 'bit' was the engine mounting and undercarriage. The draughtsman sketched out the geometry of the structure he proposed, with dimensions that fitted in with the overall concept, and this was a space frame. This is a three-dimensional lattice of struts and tension member capable of supporting the engine and taking its pull, attached to the wing structure to which it transferred the loads, and also having the struts which would form the legs of the undercarriage, with bracing to take the drag of brakes or any side forces when landing in a cross wind. The frame was entirely pin-jointed, in other words free-moving connections to the adjacent members of the framework so that all forces were axial tensions or compressions. In structural terms it was 'simple' rather than 'redundant' in that there were no alternative paths for the loads to be taken. My job was to set up all the equations to resolve the forces and to prepare the detailed calculations that would have to be carried out to cover the full range of possible conditions. This was the engine weight, the weight of the whole plane and its load on the undercarriage, full power, full braking, the shock of a heavy landing and manoeuvring in the air. These were generally interpreted as extra gravity, i.e. g-forces and there were standards laid down, based on experience, for what multiplier to apply to the basic static loads. This was really vector algebra in setting up the tables of calculations covering all possible cases for the girls in the computation department to carry out. These were what are now called spreadsheets but this was before the days of electronic computers. If they had existed I would have just had to put the data in, push a key or two and out the answers would come! Then the girls had Swiss electromechanical calculators that could add, subtract or divide. They then had to feed the calculations in, box by box, column by column and line by line, and send it back when complete several days later (where and how these state-of-the-art Swiss machines were obtained I don't know.) The results were then given to the draughtsman, so that he could decide the sizes of alloy tubes needed to meet the maximum tension or compression in each member, and design the special castings that these members would be fastened to by bolts forming hinge pins at each junction. So was the whole design built up and the production line was set up for mass production. In due course the prototype flew and the test pilot put it through its paces. However, it was getting close to the date when the atom bombs were dropped on Hiroshima and Nagasaki and the war with Japan ended suddenly. There was not going to be the same need for transporting or supplying ground forces by air. The production line was stopped.

In fact that basic design of the Bristol 170 (Fig. 5), the Freighter with a loaded weight of about sixteen tons, was modified to civilian use as the Wayfarer, and many were built and formed the 'air bridge' between Kent and Normandy that took holiday-makers and their cars to Europe for many years before being ousted by roll-on roll-off ferries and the hovercraft. The plane's front-opening clamshell doors and over two metres height in the cabin made it very suitable. Later, these planes found service island hopping in the Pacific, where runways were short. Obviously my structural design work stood the test of very many take-offs and landings, but the responsibility of designing something that, if it failed, would probably kill the test pilot and his crew was somewhat stressful. I was only twenty-one at the time! The big job however, came as a result of Lord Brabazon's review of the requirements for civilian aircraft in the post-war period, the major recommendation being for a plane that could fly its load of passengers from London to New York, non-stop. Indeed this was given his name as the Bristol Brabazon, the largest passenger plane in the world at that time at 130 tons weight, less than half that of the present jumbos (Fig. 6). There was a large team of draughtsmen, stress-men and other design staff working on the project which was conceived as an eight-engined plane, using the new gas turbines then being developed, jet engines that would also have propellers, called turbo-props. These would be entirely inside the rather thick wing, so as to produce minimum drag, but because there was a delay in getting them into production the first four prototypes were to be fitted with conventional piston engines, again entirely within the wing and with each pair of the eight engines driving a counter-rotating propeller. This was so the airframes could be fully flight-tested before the production line went active.

However, Britain was in dire financial straits just after the war. There was so much air-raid damage to repair and new housing to be built. The arms supplied to our forces by the USA were not gifts: lend-lease meant that they would have to be paid for, and in 2006 I read somewhere that the last payment was about to be made! Government spending had to be curtailed, and the Brabazon was an expensive project. So although prototype 1 went through its flight test programme under Bill Pegg, the project got cancelled. There would have to be several return flights each day to make it pay and no one in government thought that there would be several hundred people a day wanting to cross the Atlantic non-stop. Luckily I saw the Brabazon on its fly past around the coast before being sent to the scrap heap. We then lived in a flat at Ainsdale on the Lancashire coast.

Efforts were turned to short-range planes, the Britannia 1 using the new turbo-prop engines. It could get from Shannon in Ireland to Gander in Newfoundland so London to New York took three hops, but the Americans were not standing still. They had the resources and the skill to design longer-range planes that could do that journey in two hops. So we went on to Britannia 2, then 3 then 4, all too late, too small or with too short a range to beat the American opposition. They went into faster pure jets before long and so there was a long period before any European planes could really compete in the long haul work. And of course the demand for worldwide air travel expanded rapidly. Today many thousands fly daily from the UK to America.

Under the direction of labour no one could change their job without permission, so it was some time after the end of the war before I could leave the aircraft industry to take up civil engineering proper. The experience at NPL and then Bristol did me no harm of course. It had given me responsible roles very early in my career, which would be useful as a civil engineer too.

CHAPTER 6

Civil engineering at last

I<small>T WAS IN 1946 THAT THE EMPLOYMENT</small> controls were relaxed so that I could apply for a job in civil engineering, and there were many opportunities becoming available as the nation turned its attention towards rebuilding the infrastructure after the seven years of wartime neglect and bomb damage. The first employment for a recently qualified engineer in those days was as a Graduate Assistant under Agreement, a status indicating that the employer would see that your theoretical training would be supplemented by experience in the job, being supervised and guided by experienced engineers. In effect this was a requirement of the Institution of Civil Engineers before one could apply for Associate Membership. I had been a student member since I was eighteen when I went to college, and had been able to attend meetings of that august body while I was in London. One of the nation's requirements was for new roads to meet the development of traffic and James Drake, Chief Engineer of Lancashire County Council, was a strong exponent of the need for a system of motorways. LCC were recruiting staff for the purpose and having in effect not lived at home since I was fifteen it made sense to apply, having established that I could travel to the Preston office by train from a station about a mile from my parent's home in Vaux Crescent. I applied and was successful and on 6 August 1946 I set off to become a real civil engineer!

There are incidents in one's life when you think what a strange coincidence that was. Waiting for the train that morning – old steam-hauled trains on that line – there was another smartly dressed man of about my age who got into the same compartment. 'Hello! Where are you going?'

'To Preston.'

'Oh, so am I. I'm starting a new job.'

'That's strange, so am I – at the county offices.'

'Well so am I, in the County Surveyor's Department.'

'That where I'm going, as a civil engineer.'

'So am I!' This was Eddy Naylor, and so began a lifelong friendship, that grew into family friendship and we have always kept in close touch though we now live too far apart to meet very often. Eddy's career began in the bridges section of LCC whereas I was in the highways department. This involved preparing designs for various road improvement schemes, which meant going out surveying the area, with theodolite to set up a grid of lines, chain and tape to take offsets to all features, taking levels of the surface features, inspecting the availability of drains, and then drawing the plans of the work and taking off quantities. In other words, how much excavation was needed, how much stone pitching for the foundation, how many square yards (not yet metric!) of tarmac or cubic yards of concrete, the gulleys, drain pipes, and kerbs all to be listed in the bill of quantities on which the likely contract price would be determined. However, the most interesting work was the original field surveys on which feasible lines for the M6 would be chosen. On this, I worked with a Welshman whose experience enabled him to look at the contour maps and sketch possible routes that we would drive to in his car and inspect. We had no right of entry to private land but from prominent spots one could see most of these routes. At that stage even which valley the M6 might follow was not determined, and there was a lot of difficult terrain to cross in Lancashire and into the Lake District. We would take a series of levels which we could relate back to benchmarks, and plot the longitudinal sections and get some idea of how much excavation and fill would be required. This led to 'mass haul' diagrams, indicating how one could efficiently move earth and rock from excavation zones to form embankments elsewhere. The first section of the M6 planned was referred to as the Preston Bypass, that town being quite unable to handle the amount of north/south traffic trying to squeeze through its narrow streets. All this was good experience, but as this necessary period of training was coming to the end, I had the option of looking for a new job, and the possibility of avoiding the daily commute to Preston by getting a job in my home town of Bootle seemed attractive. A job in the Borough Surveyor's Office became available and I started there in the autumn 1948.

I was quite happy to work for the council that had given me a scholarship to help me on my way to university and that had provided my excellent basic education. Also, apart from living at home still, I would have a bit more free time to meet my local friends, and especially my girlfriend, Margaret, who lived some two miles away in the McGeagh family home in Litherland. We were then twenty-four so had grown out of the youth club scene of the 17-Plus Club, but we still had the same circle of friends, of course. Bootle's priority was postwar house-building to replace the many thousands of homes destroyed by the bombing, and many more demolished due to damage and their failure to meet modern housing requirements. In fact the field which I used to walk across on my way to school was now filled with about a twenty-foot depth of bricks and building debris over the whole area of perhaps two acres! New estates were to be built on the outskirts, and additional engineers were needed to design them and supervise construction. In effect I was the resident engineer on one of these, the twenty-eight acre project for perhaps 400 new homes. This site was on a soft yellow sandstone and in effect the drainage had to run against the natural ground contours to marry up with an existing main sewer. Thus excavation in rock was needed, in places up to fifteen feet deep, and this was done mostly with back-acting excavators or drag-lines, after the soil above rock level was removed by tractors pulling scrapers. They were forming the ground level for the roadway foundations too. The rock was broken up using explosives. The drains and sewers were partly salt-glazed stoneware pipes but larger sizes were made of precast concrete. In conjunction with the contractor's engineer, my job was to see that everything was done according to the drawings, the levels were correct both of drainage and the roads. House-building would start after the site works contract had got to the stage of having all drains complete and the roads finished apart from the final surfacing. Another part of the job was 'measuring up'. This was to agree with the contractor exactly how much excavation had been in rock, for which a higher price was paid, as well as the areas of stone pitching, lengths of pipelines etc.

There was another site to be developed not far from my father's watering hole, at the Old Roan pub. This Copy Lane site was then outside the borough, because no more land was available within its boundaries. This site required its drainage to be pumped and I had the job of designing the pumping station itself. This was an ejector system, a very robust system where the sewage goes into a large cast iron pot, sealed at the top, and when this is full to a certain level an air compressor pumps air in and forces the sewage out into a rising main, then going at fairly shallow depth through pressure piping to an existing sewer. There are flap valves on the pipe connections to the pot, to allow sewage from the gravity drains to flow in, and then force it out under pressure into the rising main. The compressor was electrically driven but with a stand-by diesel engine started by turning a handle and then pushing a knob to make it compress the fuel to ignite it. Not an easy operation unless you had the arms of a wrestler!

Being back in Bootle, aged about twenty-four, seeing my girlfriend Margaret McGeagh on a regular basis, visiting each other's homes and catching the 11.09 bus from near her house, it was obvious we were serious about our joint future. We became engaged, and the wedding was fixed for 27 August 1949. We had already discovered that we were born on the same day and I don't know whether an astrologer would regard that as a good omen, but to us it seemed we were fated to share our lives together. If it was an omen, it was a good one! I was learning to drive so that I could use one the Corporation's little vans to do site visits and passed my driving test somewhere among the dockside roads of Bootle, but money was tight as we were saving towards our marriage and the prospect of setting up home. Many employers in the public sector were at that time offering rented accommodation to staff they recruited, as otherwise young professional staff could not afford to take up appointments. Bootle was one such so I enquired whether I would qualify, but the answer was 'no'; I was existing staff. So after our marriage, we lived with my parents, using the spare bedroom. But of course any young married couple turns their thoughts to starting a family and this would not really have been feasible unless we had a place of our own. In any event, for engineers in the municipal sector it was best to move on to a new job every couple of years, to earn a bit more and start working one's way up in the profession as one's experience increased. It was time to look for a new job.

My best mate from university got in touch. Would I be best man at his wedding, just a few days after our own! We planned to honeymoon in Ilfracombe in Devon. His fiancée's parents lived in Yeovil, so it was definitely possible. Our wedding was at the local church of St Philip's, an ordinary Church of England ceremony. Going to Frank Sibley's wedding was quite different, as his in-laws-to-be were high church, more like Roman Catholic with a nuptial mass. His bride Sheila was worried whether she would survive without food until the afternoon and the seemingly interminable service. Being best man, I was on the front row of pews with the bridesmaid, who nudged me as the service got going and said 'I hope you know what to do. I'm following you!' My response was to say that I was following her, so from then on it was a case of half turning to watch the row behind to know when to kneel, stand, bow, bend your head, cross yourself. We got through it. Strangely, too, our good friend Eddy Naylor asked me to be best man at their wedding, exactly a week after ours – but we would be miles away in Devon so I had to decline. Marriage was in the air, and the Naylor children and ours were born at similar times, though they went one further, and the families grew up good friends and we shared some holidays together.

Our first son, John, was born on the 8 January 1951, and by that time I had got a new job in Southport, with a flat on the seafront at Ainsdale, a few miles south of Southport itself, with the Cheshire Lines railway providing a train service into Lord Street station. These were old steam-hauled trains that took a roundabout route via the villages between Liverpool and Southport, so really not competing with the direct electrified line. The Worsicks occupied the flat below and he also was employed by Southport Corporation. In many ways it was a very pleasant place to live. There were few houses there, but directly in front were miles of sandy beach. The tide went a long way out, but came in rapidly via tidal creeks parallel with the shore line. One had to be careful not to get trapped and we saw quite a number of cars get stuck in an area of soft sand, sometimes inundated by the incoming tide. Bank Holidays seemed prone to this, as so often the weather would break late in the day. On a clear day we could see Blackpool Tower, and if very clear the Lake District and North Wales mountains – and even occasionally the Isle of Man on the far horizon. We could walk in the sandhills behind, with their natterjack toads. But it could be very bleak and windy. Sand could get everywhere in a gale, even into clothes' drawers. Occasionally we went down to the Worsicks' flat to watch their little black and white television. We had radio and record player, so could indulge in our shared enjoyment of classical music. We had a spare bedroom, so our parents, other relations and friends could come and spend some time at the seaside, in good weather sunbathing and swimming. By that time, I felt we could afford a car and got a 1933 Austin 10 tourer, with a very leaky hood! But at least we could travel to see friends and conveniently visit the families in Bootle and Litherland. One of the winters we were there was very cold, and the boating pond froze. Margaret's brother Harry came as he could skate. Coming home from Bootle one Sunday evening there was snowfall that froze onto the surface and we only just managed to get home, the wipers being inefficient in freezing snow, and no heater in this ancient car. I got the car to the front of the flat, but there was then no way it could move on the ice to get put away in the garage behind the building. It was our first home, and we greatly enjoyed re-decorating it in the latest style, acquiring our Utility furniture (we still have the dining chairs) but generally making do with what we could afford from savings. That first home is recorded in the name 'Ainsdale' that we gave this house where we now live.

During the period when in Bootle and later in Southport I was working through a correspondence course to get my Testamur of Municipal Engineering. Whilst working in that field it was very desirable to get the appropriate qualification of Member of the Institution of Municipal Engineering that required some extra knowledge not necessarily covered in one's civil engineering degree course. I had by now already got enough experience to become an Associate Member of ICE, and would later qualify as Member and finally as Fellow. In due course I also became a Member of the Institute of Public Health Engineers, of the Institution of Water Engineers and the American Society of Civil Engineers, but allowed the IMunE and PHE to lapse when they were no longer of interest or advantage to my career. Although it is over ten years since I fully retired, my Fellowship of ICE remains a valued mark of my career.

My work in Southport consisted of the roads and sewers for another housing estate, effectively as resident engineer, but the main job was the Lord Street Flood Relief Scheme. This involved a precast concrete pipeline from about 36 ins to 48 ins diameter so one could walk through the larger pipes to inspect them. Southport being on sand and not very much above sea level posed particular problems in any work involving excavation as the ground was waterlogged. The old way of digging a trench was to drive down vertical boards fitting close together side by side on the outside of a timber framework. Any water coming in would bring in sand and hence undermine the surrounding ground leading to potentially severe collapse, so any leaks had to be sealed by pushing in anything that would stop the sand getting in. Upward leaks from the floor of the excavation were reduced by driving the boards at each side well below but it was still fraught with danger so that deep excavation was out of the question. There was a new method recently introduced however, which would be ideal for this scheme: well points. This was based on lowering the whole water table in the area, by putting down vertical suction tubes several inches in diameter and spaced at about ten feet all round the site. These were all linked to a suction pump that could deal with air and water in any combination. The well points, which had a fine cylindrical screen at the lower end, were inserted by jetting them down, a very easy operation in this pure fine sand, even if there might be some gravelly layers. Once the water table was low enough conventional excavation by backacter machines was possible, but the trench sides still had to be supported by a system of steel waling (a technical term for supporting the walls either side) on a timber framework to stop them caving in, but there was no danger of the nasty phenomenon of running sand (Fig. 7). The drainage system linked up the existing surface water sewers, where they were not already full of infiltrated sand, to a pumping station on the promenade which raised the floodwater when necessary to go out to sea through a piped outfall. The pumping station and outfall pipe were different contracts. I had to look after the Lord Street end of things, with general supervision from my boss, Mr

Thornburrow. It was a big, challenging job involving the closing of the main roads through the town.

This job was in due course to bring me to my first experience of legal processes. All the work done was being surveyed and measured for comparison with the bill of quantities on which the contract price had been based, and all these measurements were agreed with the contractor's site engineer. My boss was a bit of a stickler over contract conditions – why shouldn't he be? – and although I can no longer remember what the issues were, the contractor put in a claim for extra payment amounting to a large proportion of the contract price, which would have almost doubled the cost to the Corporation. They were prepared to meet some part of it, but not all so the contractor took to the law, in effect, electing to go to arbitration. This meant putting the case for and against the claim to an independent arbitrator appointed by the Institution of Civil Engineers, and this was John Calvert, a distinguished consulting engineer with lots of experience in drainage matters. I had left Southport by then and the case was heard at the ICE, in Great George Street, London. It meant all parties spending perhaps a week there. There were QCs appointed by both sides, so there was quite a bit of legal argument. The contractor's main claim was that he had been put to extra work by the requirements of the engineer, Mr Thornburrow, and produced what were called day-work sheets to support this. There was one for every day of the work, for problems of delay through traffic diversion etc. He was supposed to have got any such day-work items outside the contract agreed as we went along, day by day but never had. They were all on his standard form, each dated. I noticed that they were all, without exception, dated a year ahead of when the work was done. Were they contemporary records as claimed or were they all made out much later to support the claim, which to our side seemed mostly totally spurious? This anomaly was pointed out to our QC but he said we must not mention our suspicion in court as it was not privileged in the legal sense, so the other side might claim for defamation if there was any suggestion of malpractice. The only ones who really won were the lawyers! The contractor got half his claim, Southport Corporation saved half in theory but the legal fees swallowed up the difference! Was justice done, I wonder. I used to meet John Calvert regularly many years later when I was with Binnie & Partners because his firm had offices in the same building. Did he ever remember me from those earlier days? He always called me Peter if we met on the stairs, but he was of an older generation so he was always Mr Calvert to me.

Another feature of work in the Town Hall on Lord Street was the pleasantries in the drawing office. In those days an engineering department was in fact a drawing office because so much time was spent doing the plots of surveys, adding the plan of the scheme in mind to the plan and then preparing details such as large scale cross-sections and the arrangement of steel bars in reinforced concrete. Calculations were usually done by slide rule, with five- or even seven-figure trigonometric and logarithm tables for working out such things as theodolite traverses. So we were a bunch of chaps of various ages working at our drawing boards, with plan chests of drawings of previous schemes and Ordnance Survey maps of the area. One of the engineers was a great wag and for some reason often turned his tricks against one man in particular, about ten years older than me. He was the first to learn about special hard drills coming on the market that would drill into brickwork. So poor Steve Laycock found his coffee cup screwed down to the bench top one break. Another time there was a kipper fastened to the underside of the drawer in which he kept his pens and pencils, but his *pièce de résistance* went rather over the top as it frightened Steve's wife. He was in the habit on a Friday morning of going to one of his sites on the outskirts from where he could nip into the country to buy a dozen eggs at a farm. He brought the box into the office but one week he also had to go out to site in the afternoon so the office wag painted the whole boxful of eggs with black Indian ink! His wife got up on Sunday morning to prepare the bacon and eggs and almost passed out on seeing the jet black eggs when she opened the box. They did look awful and Steve was none too pleased when he got to the office on Monday.

I was getting into serious black and white photography at that time, and the player of tricks was also a chap who picked up army surplus goods, and would pass items to me that might help me in making my enlarger, so he was not always mischievous.

In those days, in municipal engineering it was desirable to move on every few years to get promotion and widen experience so in

1952 I applied for, and obtained, the post of Senior Engineer in the Main Drainage Department of Stoke-on-Trent. We then had a council house at 65 Oliver Road, Harpfields, and Margaret was pregnant with our daughter Sheila. This was a new estate, a bit out of town, and on the heavy clay soil which characterises the Potteries. The city was old and its sewerage system was old (Fig. 8). Hence the need for an entirely new system of main drainage. The first requirement was to find out what was already there so the initial task was to prepare sewer maps that would show the existing locations, sizes and levels of the old arrangements. With much new development going on they were already overloaded and the convention then was for excess flow caused by rainfall to be discharged through overflows, usually weirs at the side of the sewer over which the excess could escape into a stream or ditch. Too frequent overflowing meant very nasty pollution so there were many complaints to spur the city council into action. I had another engineer perhaps a year or two younger, John Hall, who joined me to form a survey team, and there were other teams covering other parts of the city. We usually had a sewer-man with us, a workman with pickaxe, sledgehammer and lifting keys to open manholes. We would lower a lamp down first to check whether there was any noxious gas down there before one of us would climb down to measure the sewer diameter and note the direction of flow and any branches coming in, which would all get followed in turn to build up the full map of what lay below ground. We carried a surveyor's level, so could take the ground level in relation to an adjacent Ordnance Survey point – no global positioning and levelling in those days to make it easy – and depth to the invert of the pipe. Some of these manholes had not been opened, and certainly not entered, for perhaps fifty years, and although most had step-irons built into the side of the shaft, they were not always secure. On one occasion, going down one of these, a step-iron gave way under my weight and down I went. On the way down, I thought, keep your chin back because if that catches on one of the step-irons below you could break your neck. Better to break a leg. But at the bottom there was a couple of feet of dense sludge to break my fall so I escaped without injury!

The design of drainage systems which take surface water as well as sewage first requires an estimate of the maximum flow. The foul sewage content basically comes from the water consumption per household times the number of houses in the area that any drain serves. The rainfall component is the product of the maximum rate of rainfall to be allowed for, times the area on which it falls, expressed as the equivalent impermeable area. For urban schemes the convention then was the heaviest shower of any particular duration likely to occur on average once a year (the present day standards are rather higher especially in city areas). The duration used, anything from five minutes to perhaps an hour, is taken to correspond to the length of time the flow would take to get from the head of the catchment area to the point being examined, which defines the shortest most intensive storm that needs to be considered. The design of the drain pipe itself, its diameter in relation to the available gradient, is worked out from a flow equation, a formula that describes how much flow will go along a pipe under a known pressure gradient. Hence a network of design data is built up for the proposed system.

The foul sewage gets taken to a treatment plant, probably on the outskirts of the town and a new one was proposed for Stoke, but all the rainfall on the area cannot be treated in the same way: there is too much of it! So it conventionally gets overflowed, and in those days weirs at the side of the sewer were typical for this purpose. The design of these was done according to a standard method but my knowledge of hydraulics soon told me that the standard method was totally wrong. This was one of the topics that took up my spare time, looking at the available research in the published literature and trying thus to produce a design theory that was in accordance with hydrodynamic principles. I worked on this using published experimental data and in due course submitted a thesis on it to my old college for my MSc(Eng), and this later formed the subject of a paper published by the Institution of Civil Engineers and presented at a meeting at 1 Great George Street with three other papers on the subject of storm overflows. Another topic that was much under discussion in the professional literature then was how best to perform those computations for the relationship of run-off to rainfall in a catchment, with a largely graphical method emerging, so there was a lot of interest, discussion in the office, correspondence in journals etc. on this and similar issues. The new breed of civil engineers wanted to use up-to-date knowledge and best available methods rather than the pre-war procedures that had been inherited from the Victorian engineers. It was this that got me very interested in research and its application in practice.

In any new drainage system, there is a fair chance that the level difference between the property served and the land where the treatment plant is sited does not allow a gravity flow connection, and if so a pumping station is required. This was so in this case and one of my jobs was to design the pumping station, quite a large one of similar scale to the one on the Lord Street flood relief scheme in Southport. This was to be at the low point of the area, obviously, in the district of Hanley, surrounded by potteries and the mills where the clay was prepared and bones ground for the bone china being made. But gradually these old coal-fired pot kilns were dying out and new works were being built outside the city so there was a lot of dereliction in the area. One of the principles of designing pumping stations in urban areas is that there must be no connection between the wet well, where the flow enters, and the dry well, where the actual pumps are, other than the suction pipes leading direct to the pump. This is because of the risk of gas, and sewer gas, very similar to methane, marsh gas and the firedamp found in coalmines, is highly explosive. It will rip a structure apart, kill anyone there, if there is a spark to ignite it, even just from a conventional light switch. Much later in life I became only too well aware of the seriousness of this risk if it is not rigidly guarded against without compromise.

With my increasing interest in research topics, especially any to do with water, it was natural to apply for a post that was advertised in 1956 for a Senior Scientific Officer at the Hydraulics Research Station, Wallingford, then part of the Department of Scientific and Industrial Research. By that time, our third child, David, had been born, with some considerable difficulty during labour that effectively put an end to any further extension of our family. I prepared for the interview at Wallingford by making sure I knew what the principles of hydraulics were – I had been well taught at university anyway – so had no problem getting the job. The town of Wallingford was very pleasant, connected to the main GWR line by a little steam train running along a branch of a couple of miles. The laboratory had only been in existence at its new site for a couple of years, the organisation having been set up at NPL near London. It was about to expand, so why not get in at the ground floor? A new experimental hall was near completion, but the headquarters were in an attractive old manor house. Housing was available, so it all slotted into place. Wallingford would be a nice place to bring up a family as well as to work.

We had found Stoke-on-Trent pleasant enough and convenient for visiting my parents, still at 4 Vaux Crescent and Margaret's parents in Hatton Hill; also my sister, who lived in Connah's Quay, North Wales, with her schoolmaster husband, Cecil, ex-Squadron Leader, C E Humphreys, with a distinguished flying record for service in the Mediterranean theatre during WW II, but the time had come for my career to take a new turn, leaving municipal engineering after about ten years and going instead into scientific research.

CHAPTER 7

Specialisation in hydraulics research

T HE CHANGE FROM PRACTICAL CIVIL engineering to research was
not a particularly great change because that being carried out
at the Hydraulics Research Station (HRS) was very much applied,
with direct relevance to practical problems in the short term. It was
very different therefore from purely academic research as a thirst
for knowledge whether any direct use of that knowledge was
foreseen in the near future or not. Academic research was
essentially the preserve of universities, although there was overlap
in our activities. HRS did a great deal of commissioned work,
where a client with a problem in hydraulic engineering would pay
for us to investigate it and come up with a solution. Often this
would be by using scale models through which water would be
pumped, and the rules governing this procedure are well established – the laws of similarity. The field covered was the whole of
hydraulic engineering: flood protection schemes; spillways from
dams; flow measurement; ports and harbours; irrigation systems;
tides and estuaries; waves and beaches; and sediment transport
problems. Not all work was directed towards a particular problem
in one location; some was concerned with more general problems
where improved design methods were required.

HRS was organised in three separate technical divisions, which
tended to cover structures and rivers, estuaries and coasts, and then
all the services needed such as skilled model makers, electricians,
instrument makers, experts in control systems and instrumentation,
and the more mundane administration and library. The headquarters were in an old manor house, but the laboratories were
newly-built experimental halls, the main hall as it was called having
seven bays each 100 feet square unobstructed by pillars so there
would be large spaces available for model building. Many of these
models were of tidal problems and HRS really led the world in
terms of its capability for carrying out such work. Being a new
set-up meant that there were good prospects for promotion as it

expanded, and in due course I was a section leader and then a division head and assistant director. I was there a total of sixteen years, before moving into consultancy. It is therefore not feasible to detail all that I and my division did, but the list of research papers appended shows the subjects covered there between 1956 and 1972. (Some later papers also derived from work at HRS.) Most of the commissioned studies involved model-testing and the reports on these were prepared for the clients, and were therefore unpublished. Just a few highlights from the world of hydraulics research are therefore covered in what follows, generally dealing with generic research. One feature of this field of work was the contact made with other researchers throughout the world and there was – and still is – an International Association for Hydraulic Research (IAHR) which I joined, attended its conferences, got known internationally, and was later (after leaving HRS) appointed to the IAHR Council for a three-year stint. In due course I was awarded with honorary life membership 'for services to hydraulics research', an award which I have treasured.

HRS was a new organisation, part of the Department of Scientific and Industrial Research. Its Director was Sir Claude Inglis, ex-Indian Civil Service, who had been Director of the Irrigation Research Institute at Poona, so came with excellent credentials in hydraulics, and especially in the stability of rivers and unlined irrigation canals and the control of sediment in rivers. These were to be significant aspects of the research at Wallingford. Another ex-India man was Ted Crump, who had great intuition for applying hydraulic theory to real problems. Another ex-India hydraulician was A R Thomas, consultant to Binnie & Partners whose position I was to take over sixteen years later when he retired. There was a lot of freedom in those days for HRS to determine its own research programme and one item on the agenda was to make the application of the latest knowledge about flow resistance in pipes and channels easier to apply – specifically the Colebrook and White formula. This expressed how the resistance to flow varied with the viscosity of the fluid and the velocity of flow and, like all sound equations in hydraulics, was expressed as relationships between groups of variables, assembled so that the combination of units in each parameter (diameter, pressure

1. *Bomb damage, Rimrose Road, Bootle, 1942*

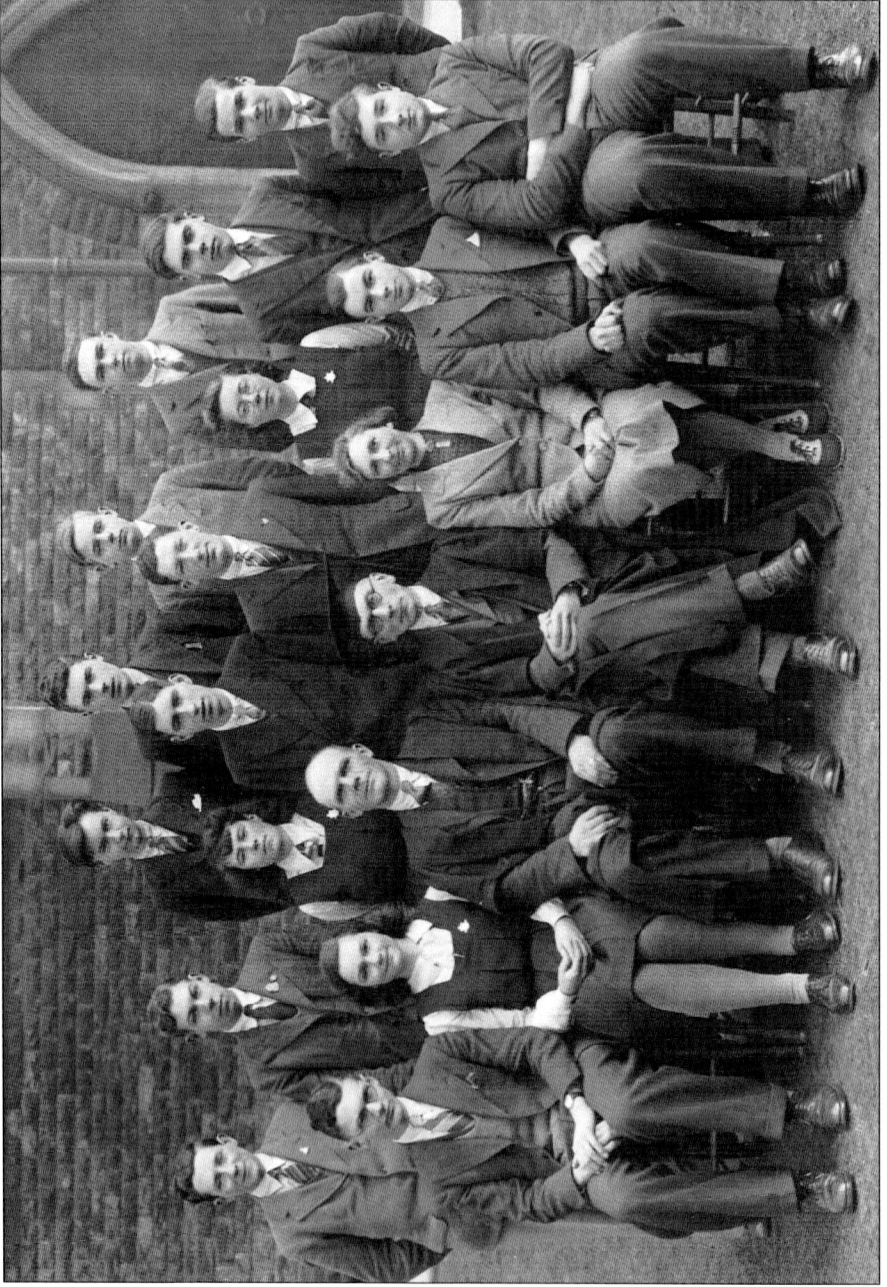

2. *The sixth form, Bootle Secondary School, 1941–1942. I am at the far left of the front row*

3. The 17-Plus Club on a coach trip

4. *Medals: Coopers Hill Medal, Henrici Medal for Mathematics, Unwin Medal, Institution of Municipal Engineers Medal*

5. The Bristol Freighter, type 170 (reproduced by permission of the Aeroplane magazine, www.aeroplanemonthly.com)

6. The Bristol Brabazon

7. *The Lord Street Flood Relief Scheme under construction*

8. *A corner of Stoke-on-Trent, when I was a member of the Photographic Society*

9. *An experimental meandering channel at Hydraulics Research, Wallingford*

10. Flip bucket spillway at the RSK dam

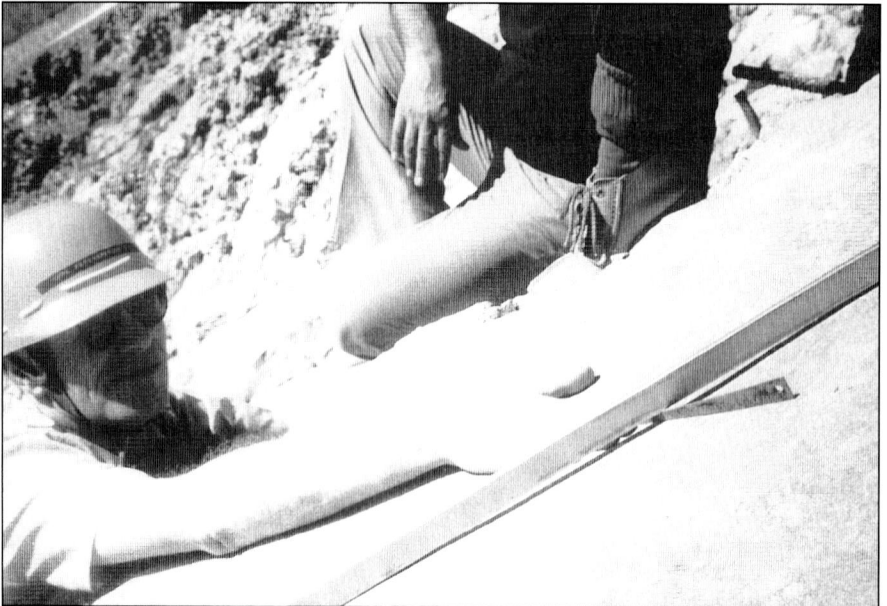
11. Checking the surface of the RSK dam. Severe cavitation in the background

gradient, velocity, gravity and fluid viscosity) came to zero – they were numbers. However, these numbers mixed up the variables the engineer wanted to use and so they were difficult to apply in practice. My first job was to solve this problem, and I found that the best place to do some deep thinking about knotty problems was in the evening in a hot bath. Like the tale about Archimedes, there was a sudden flash of inspiration: eureka! – I could solve it! When the solution came to me I could not quite understand why no one had thought of the simple principle before: it was to recombine those dimensionless groups so that none included more than one of the engineer's parameters, of slope (or pressure gradient), velocity or size. It was then possible to prepare sets of design graphs for different roughnesses (a linear measure of surface texture) and for a given fluid. These charts were published and went to many editions, but were overtaken before very many years when the new computers came to our assistance and so sets of tables could be prepared, and typeset directly from the computer output. These tables are still very much in use, and have gone through many editions, even giving rise to a pirated edition in Hong Kong.

Sir Claude also set me and my small team the job of establishing the area of sand which he had installed, called the meander area, as a research facility (Fig. 9). The first task was obviously to generate small meandered channels which could then be measured in several ways to study their characteristics, and why some channels meandered and others did not. We succeeded despite that period being one of the coldest in recent years, with temperatures staying below zero twenty-four hours a day for almost three months in early 1963. By this time we had saved enough for the deposit on our new house in Moulsford, a few miles from Howbery Park. The children could build an igloo from the frozen snow and there was much pushing of cars to get them started in the morning, when cars could be difficult beasts. By that time we had our first new car, a bright red Ford Popular, which surprised our neighbours because up till then cars had been any colour as long as it was black or equally sober!

Being so much involved with research, it was appropriate for me to join the body representing the hydraulic sciences at world level, IAHR, the International Association for Hydraulic Research. The first of the biennial congresses that I went to was in Leningrad in 1963, and although I don't recollect much of the technical content of the papers presented there apart from my own about model studies we had done to help design training works at the site of a new bridge over the River Kaduna in Nigeria, there were other experiences that have stuck in my mind. This was not long after the great train robbery, when thugs stole a huge amount from a train they ambushed in Buckinghamshire. Strangely, this was a topic of conversation for English language students in Leningrad, but it was still a time of repression in Russia. Several of them stopped to talk to us, well away from buildings when we were out walking after the meetings one evening, with lookouts posted around looking outwards to check there were no secret policemen around to overhear them. What they were doing was illegal! Their favourite author was Conan Doyle and the Sherlock Holmes books. One of our group was asked how much he wanted for his shoes! One of the most moving experiences on that trip was going to the memorial for the countless thousands who died there during the siege by the Germans in the Second World War. For the tour after the congress proper, we took a sleeper train to Moscow, so saw all the sights there, before flying down to Socchi on the Black Sea coast, a holiday resort for the party faithfuls, where we visited a laboratory dealing with coastal problems. We went from there to Tbilisi, in the Caucasus, midway between the Black Sea and Caspian Sea, which looked a very pleasant place. From there to Ercvan in Armenia, where again we were accosted by locals wanting to practise their English. One had come back after having emigrated to America in the belief that they would have a good life there but said they were misled. His prize possession was his American trilby hat, which he still wore.

Another early overseas trip was to America to see what the hydraulics laboratories over there were doing. This involved going south to Vicksburg, where the main research centre for the USBR, the US Bureau of Reclamation, was, and then over to the west coast to San Francisco and the University at Berkeley. The trip south involved a change of planes onto a rather modest and elderly plane, and, being on my own and having my boarding card, I was happy to get on via the end of the queue, even though seats had not been allocated. That was my first experience of the racial segregation in the deep south. There were six or seven blacks on the flight and I let them go first as there were family groups, but when I followed them on to the plane, being last on, I realised why they were a little surprised at being given priority. They had to sit at the very back, and therefore so did I!

As my team grew with the expansion of the lab. there were new fields of generic research, one of which was flow measurement. using weirs and flumes. It was then that Ted Crump was advocating a new form of weir profile and one of my team, Rodney White, did much of the detailed testing. This field of research led us, some years later, to publish a manual on *Weirs and Flumes for Flow Measurement*, including as authors others of my team, Tony Harrison and Alan Perkins. I was fortunate in having some clever chaps working in my section so that we formed a very effective team. Another of our fields was collaborating with the Road Research Laboratory in a programme of research into the relationship between rainfall and run-off in urban areas, which required the help of the instrument makers of HRS in developing methods of measuring and recording the flow in sewers and storm drains when there was a downpour. Of course this related closely to my municipal engineering experience. Another such field was storm overflows from combined sewers: we carried out a pro-gramme of model tests on various designs using sediment and dye to represent various types of polluting material.

Alongside this generic research, there was a continual pro-gramme of model studies of particular problems, when some radical thinking was needed. One tidal problem involved wind stress, another the representation of tug forces on a manoeuvring large ship, another the possible vibration of a siphon structure. In hydraulic structures such as spillways from dams it was often necessary to examine the efficiency of energy dissipation, the potential for scour and the possibility of low enough pressures at the solid surfaces as to cause them to cavitate. Cavitation occurs when the pressure tries to get below the vapour pressure of water, so that pockets of vapour form which collapse further downstream and can pluck the concrete from the surface. It can be very destructive. There was a range of clients for these model studies:

water authorities, river authorities, harbour authorities, hydro-
electric generators, and the Central Electricity Board concerned
with cooling water intakes. It was quite normal to visit the sites,
of course, to make certain one was aware of any aspects not
obvious from surveys, but we did not have any direct role that
would involve visiting the completed work, other than personal
interest if one was in the area. One such site visit was to New
Zealand, my first long haul flight which in those days took three
days or so, in a Boeing 707. Stops were involved in Cairo, the
Gulf, Pakistan, Singapore, and Sydney on the way to Auckland.
My first sight of Australia was as dawn broke over the desert which
seemed to stretch forever! This was at the time of the Cuban
missile crisis when there was a stand-off between America and
Russia, which had dispatched ships carrying missiles capable of
delivering nuclear warheads into the heart of America. President
Kennedy would not allow it and nuclear war was a distinct
possibility when I left England and my family. Would I ever see
them again? It was an emotional journey, but on arrival at the
other side of the world the crisis was over, as the Russian ships had
turned back. I met up with old family friends from my early
childhood, the Dibbles, in Auckland and the general attitude in
that far off place was 'What crisis? It didn't seem to affect us!' It
was the Waikato Valley Authority who wanted our advice, and in
due course a model study, of proposed flood control works,
involving the diversion of flood waters into a swamp area over the
peak of the flood and its controlled release later. There was the
tail-end of a flood causing minor flooding on the dirt roads when
I went to inspect the region with the Chief Engineer. We visited
Lake Taupo, and that was the only time I have had a professional
discussion naked – in the hot spring sauna there!

Being involved in research meant keeping abreast of the
hydraulics literature, and I therefore joined the American Society
of Civil Engineers, whose hydraulics journal was the leading
publication on the subject in the world. One of the topics that was
important at that time, both in terms of basic understanding and in
applying any advances to modelling was the transport of sediment
by flowing water. How did this relate to all the variables such as
channel depth, gradient, velocity and the range of sediment sizes?

There were many equations available which would predict widely different values for the transport rate depending upon which formula you favoured. A great deal of data had been collected from experiments in laboratory flumes but it seemed as if researchers were not interpreting them in such a way as to give convincing or even compatible predictive functions. We carried out experiments at Wallingford on models having mobile beds of sand or fine gravel and laws of similarity told us what sediment to use for a given model scale in certain clear-cut cases, but there was no way that problems in rivers and estuaries with fine sand beds could be modelled with certainty. There was the possibility of using artificial materials with specific gravity much below that of natural sand, but this was very risky as was shown by a model of Morecambe Bay, which used chopped up cubes of wood about 1 or 2 mm in size. Not to put too fine a point on it, it was a disaster – but luckily it was the responsibility of one of the other divisions.

Although there was little direct funding for research on this vexed topic, I set out to get my brain around it, building on the theoretical work and laboratory experiments other people had done. This is always wise, of course, because there had been giants before me on whose shoulders I hoped to stand! To ignore their work would have been foolish. Another hot bath was called for to think the problem through: another eureka! The problem boiled down to combining theoretical concepts applicable to fine sediments, which travel largely in suspension, with those applicable to coarse sediments which travel largely along the bed, in essence seeking a transition between them that would cover intermediate sizes. It occurred to me that one might combine the two theories at the limits by putting them into such a form that one could multiply them together, one raised to the power n and the other to the power $1-n$, with n being a function of some number expressing sediment size. This would be a relation between the gravitational effect and effect of fluid viscosity on an individual particle. Whether it would work required the analysis of all published research data within this new framework. My clever man for this was Dr Rodney White, assisted by Charles Robson of the computing section, who consulted experts at Harwell on how to go about the analysis when there were a lot of variables. It worked,

and we obtained new transport functions that fitted all the data
quite well, about two-thirds of all data points coming within a
factor of two of the best-fit equation. This was published (though
not without difficulty in getting it through the reviewing process)
in the *Journal of ASCE*, and got worldwide recognition as a good
predictive method, the Ackers-White formula. It stood the test of
time well, but by degrees over the last thirty-five years since that
work was done other research has led to equations that perform
somewhat better.

During the period 1956 to 1972, there was great development
in computing and in computational models. At first, HRS just had
one or two mechanical calculators, the same Brunsviga that I had
used at Bristol, and one or two spiral slide rules that could be read
to four or five digits, and the staff's personal slide rules, for the
calculations we were carrying out virtually every day. The first
electronic pocket calculator I saw was in about 1960 when an
Australian academic, having passed through Singapore on his way
to a conference in London, had picked up a Japanese calculator for
about £35, I think, which would add, subtract, multiply and
divide. The shape of things to come! Also big computers were
being developed, the American firm IBM competing with the
British firm ICL to get them on the market. In those days they
were big beasts, requiring a special air-conditioned building, with
large discs and huge magnetic tape spools to contain programs and
data. If HRS was to stay in the forefront, it would have to move
with the times and install one, especially as there was much
development at universities of methods of solving complex
functions describing processes varying in time, such as tides in
estuaries. Those numerical models were referred to as one
dimensional, although this was not very descriptive. All it meant
was they could not represent two dimensions in plan, but they
could properly represent the cross-section of flow and of course
time is a dimension as well as length. The first mainframe computer
– which is what those giants of that era were called – cost
something of the order of £1,000,000, but development was so fast
it wasn't long before it was getting outdated. Programs were
produced on punched cards, perhaps hundreds of them, so woe
betide anyone who dropped a box full. There was a card reader to

transfer the program to the computer, and more cards with the input data on. Output was on a line printer that would produce masses of figures giving the results of the computation. Specialist computer staff were needed to handle the beast, and young women seemed to have all the necessary confidence to learn this new trade. This computer department came within my division, and was led by a far-sighted engineer, Charles Robson. He could foresee many of the developments we now take for granted, such as home-working using fast communication systems to keep in touch with the office, and screens which could show you the results in graphical form. Now, the PC on your desk at home does so much more, so much more effectively, than was possible in 1972.

During the sixteen years I was at HRS, the station expanded to its maximum size. It changed the government department it was answerable to on several occasions, at one time part of the Ministry of Technology but by 1972 in the Department of the Environment. Originally the special topic of hydrology, in practice the whole area of rainfall and the resultant run-off in catchments, and the understanding of rainfall data and its probability, was part of HRS remit but that was hived off to form a new establishment on adjacent land with a fine new building, the Institute of Hydrology. But by 1972, the government was trying to get rid of many of its research functions by privatising those (rather small) elements that it thought could form viable commercial operations. As HRS did much research paid for by the civil engineering industry it was an obvious client for privatisation. By that time too, I was beginning to think my role there was getting superfluous. There was no prospect of further promotion. The lab. had had Fergus Allen as Director for some years and when he moved on to a job in Whitehall, Robert Russell had taken his place – he had previously been his Assistant Director. Robert was about my age and even if he left in due course there was no chance of me, getting towards fifty already, being his replacement. I would not want to go to some other senior job within government science and become an administrator. I was an engineer not a pen-pusher, and I wanted to keep my brain active. Having rounded off my research on sediment transport, and fearing that with privatisation the scope for choosing one's own field of research would disappear completely, the time was perhaps ripe for another change of career direction. One of our very good clients was the consulting firm, Binnie & Partners, the same firm as the Binnie, Deacon and Gourley, one of the partners of which had been the close relation of my applied maths teacher. I had established good relations with them, through model studies of siphon spillways, large vortex drop structures and spillways from dams. One of the partners suggested I might like to join them as hydraulics consultant, to replace A R Thomas who was approaching retirement age. One needed permission to leave the senior Civil Service to join a private company that might make use of your knowledge and experience, but fortunately permission was granted and I started my new job in Artillery Row, off Victoria Street, London, in July 1972. I thus became a commuter, because we were happy in Moulsford, a wonderful place to live and bring up a family. They were well advanced in their school careers and it would have done them no good for them to have to change schools. So from then on I was a self-employed person, with a retainer from Binnies that effectively guaranteed me an income that would not be less than I enjoyed from the Civil Service.

CHAPTER 8

The role of innovation and research in engineering

WHEN I JOINED BINNIE & PARTNERS as their main hydraulics expert it was because the partnership recognised that without the most up-to-date methodology they would have increasing difficulty in selling their services in a very competitive market. Having been in the research business, though on its applied side, I was obviously in a position to ensure that they were bang up-to-date when designing their stock-in-trade, anything involving water, whether dams, reservoirs, tidal systems, irrigation schemes, water supply projects, drainage schemes, flood protection, coastal works, outfalls, power stations with cooling water intakes and outfalls, sewerage purification plants, and water works in general. In all these a good knowledge of basic hydraulic principles was essential and many of them required knowledge of the most recent research results. Although I had not been much involved with waves and their effects on coasts whilst at HRS, I became much involved with several projects in that field, and had to follow the established design methods and keep pace with that field of research from then on.

The results of published research would often be needed to solve the particular problems in new projects. For example, many sewage schemes were then being upgraded as coastal resorts and towns realised that their sewage outfall was polluting the beaches and neighbouring coastline. There was a call to replace any short outfalls by much longer ones going into deep water, with multiple outlets at the sea bed which would encourage the mixing of the partly treated sewage with the surrounding denser seawater, so dispersing it to be carried away by tidal currents. This was also an area for research in many countries into just how this system of buoyant plumes would mix and expand. This was one of the areas of work Binnies undertook, so one had not only to be aware of this background research, arrangements had also to be made to collect the data about tides and currents by organising field survey work.

In coastal protection and related work, there was a tendency to go into deeper waters, or more exposed areas. Protection had to be effective in rare storms, so there was a need to collect statistics about the wave climate, or to predict the worst waves to be expected in perhaps events likely only once in several hundred years. Methods of doing this were continually being improved, as well as new protection systems coming onto the market involving precast concrete blocks, some weighing several tonnes each.

Urban flood drainage schemes often involved deep tunnels, with shafts to carry water down from the shallower primary system, and this could be done best with vortex structures that ensured that the flow spiralled down the shaft with an air core. This system was also used in the Hong Kong water supply system to take flood flows from natural streams down to collection tunnels and these were large structures, designed on the basis of equations developed by Ted Crump at HRS when I was there.

Research came into the consultancy world also where it was necessary to commission model studies, for example of the spillways from dams that were proposed, or where research was an integral part of some major project. The study of tidal power in the Severn estuary was such a case, where B & P were appointed as the lead consultants with the role of advising the Severn Tidal Power Committee chaired by Sir Herman Bondi on the organisation and co-ordination of the wide-ranging studies required into all aspects of any scheme. These would involve its optimisation, the necessary configuration, its output, the effect on tides and floods, slope protection of any embankments, where any sediment would deposit, the general environment such as salinity, navigation, what locks would be needed, and the effect on ports, sources of materials, methods of construction, the use and design of caissons and how they would be towed and sunk in place, and every aspect of the impact of construction and the operation of the scheme on the whole area, not to mention how the output would be fed into the grid and whether it could indeed fit economically into the nation's energy supply system. Also included in the programme was field work on such aspects as wave climate and the variation of sediment concentration through the cycle from spring to neap tides. This was perhaps the largest project of all those that I was involved with, including drafting those sections of the committee's report to government that concerned hydraulics.

At a more mundane level, but nevertheless very interesting, were a couple of projects we carried out for CIRIA, the Construction Industry Research and Information Association. One of these was on the flow resistance of tunnels made from precast concrete linings. This project was carried out on a section of the major water ring main near Datchet, where there was access at adjacent shafts so that we could carry out accurate level difference measurements over a range of flows. Another was on the resistance of slopes protected by randomly placed rocks to wave attack. This was carried out at an artificial bank constructed in the Wash estuary. The approach was to put in a sequence of different sizes of rock on adjacent sections of the bank, to measure and record the waves there and to watch and wait! When there was a storm, the surface of the bank was re-surveyed accurately to discover whether any rock had been displaced, and although we hit a period of relatively calm weather, over a period of a few years we did indeed see significant damage on the lightest two of the rock sizes, so were able to draw conclusions for future design.

Innovation was continually being called for in many of the projects being designed, and some of this arose because of the increasing availability and use of computational models. I was very fortunate to have in my small team some excellent young engineers who had been through university when they were keen to teach computational methods and indeed some had done research at university using such methods, becoming skilled programmers. Graham Thompson was one, and he had the rare ability to adapt commercially available programs to fit our special needs, correcting any mistakes in them on the way! How many people these days just blindly apply programs bought off the shelf without really knowing whether they are accurate, and perhaps not even bothering to find out what equations they were applying! Others in my team were ready and willing to develop new programs to apply to special problems. One such program utilised the latest research on cavitation in high velocity flows to predict whether, and if so where, cavitation might occur on the down slopes of chute spillways on dams. I was able to formulate the hydraulic solution in the sense of assembling the equations and showing how they could be solved metre by metre down the spillway slope, to see whether there was danger of cavitation or not. We were the first team in the world to have such methodology, I believe, and our philosophy was always to identify what could be done in the time and within the budget that was available, unlike many of the public sector computer systems you read about today where there is gross overspending and massive delay as the so-called specialists bite off more than they can chew.

Another case of innovation – or was it lateral thinking? – was in the analysis of the probability of extreme surge affected tides. B & P were involved in studies of the possibility of a new London airport being built on the Essex coast at Maplin Sands. This would use land reclaimed from below high water level and it would have to be protected against inundation by extreme tides. The convention prior to our work was to examine tidal records to extract the highest tides observed each single year, and to analyse this series of annual highest observed tides in the context of statistical theory to forecast what might occur in a period of up to 1,000 years. There were more years of record at nearby Southend than at most other places, but even so extrapolating from them to 1,000 years seemed to be using only some of the available data. What about those tides affected by surge where the surge did not occur at around high water, or when it was a neap tide? They were being ignored by the conventional analysis, and should not be. The astronomic tides are fully predictable of course, the moon and sun and their alignment being the main influences, but there are other astro-nomic effects as well which give about a twenty-year cycle before they repeat. Tide tables give all this data. Tides are thus controlled by quite different processes from surges, which result from atmospheric pressure and wind stress on the sea surface. Tides and surges are thus fundamentally independent processes and can be treated as such in any statistical analysis: if you know the probability of each separately, the probability of their combination is obtained by multiplication. A 1 in 100 event in one might combine with a

1 in 100 event in the other to become a 1 in 10,000 event. This is referred to as a cross probability. So we analysed the tidal data by separating out the surge effect from the predicted tides, worked out their statistics and recombined them with the long-term probability of predicted tides. No one had apparently thought of doing this before, surprising though it might seem, and the new method was presented (jointly with my colleague, David Ruxton) at the coastal engineering conference in Copenhagen in June 1974. David Pugh of the Institute of Oceanographic Sciences at Bidston was there and spoke to me about this approach. He must then have drawn it to the attention of his colleagues at Bidston, who did further research to refine it. It has since been the standard methodology when studying extreme levels arising from meteorological surges.

Innovation often provided solutions to tricky little problems, such as turning flow round corners. Earlier when at HRS, in designing a flume which formed a horizontal loop, providing conventional curved bends would have generated objectionable secondary currents and super-elevation. Why not follow the example from wind-tunnel design and use cascade bends? These are sharp right angles with a series of curved fins across the diagonal which divide the flow into slices each of which is turned by the flanking vanes. It worked well when the flume was built, and the same idea was recommended and used in a number of overseas water treatment plants where it saved space and construction costs. It was also possible to use systems related to the cascade bend in turning fast flows in spillways from dams, though these were subjected to model testing too.

Innovation must not be confused with risk-taking. There is no room in engineering practice for taking risks. Innovation might mean just looking at things from a different angle, applying some lateral thinking or just being more up-to-date than any methods to be found in text books or design manuals. Applying the latest knowledge is surely best engineering practice, though one must always be wary of blindly applying a function given in a research paper to describe some research results to practical circumstances that may be outside its established range of validity. There may be temptation from clients to save money by cutting corners on projects – or even excessive pressure to meet environmental or architectural aspirations – but this should be resisted if the security of design would be compromised, because without doubt the engineer will be singled out to take the blame if the project proves to be defective in any way. With the increasing tendency to seek redress through the courts that we seem to have taken over from the American culture, innovation might well become stifled. Increasingly, the philosophy might become 'If it's not in an internationally approved design code, don't do it'! It seems to me that that would be a sad day for engineering.

CHAPTER 9

North Africa and the Middle East

M UCH OF WHAT HAS BEEN WRITTEN so far was in approximate
date order but what follows is more like a travelogue, though
with reference to the engineering that led to journeys to many parts
of the world, sometimes to very remote areas, sometimes with
slightly hair-raising experiences on the way. So the following
chapters are arranged geographically to indicate how a career in civil
engineering gave marvellous opportunities for world travel, in the
days when rapid air travel became the normal way of covering large
distances and before the world began to divide itself into two
opposed camps, Muslim and Christian/Judaism. North Africa and
the Middle East form much of the Muslim world and travel there
was fairly safe as long as foreigners respected the local customs. An
example of that was a visit to Algeria, during my semi–retirement
period when I was helping firms other than Binnies with hydraulic
problems. The project in Algeria was the possible effect of
reclaiming an area of land on the flood plain of a river, in order to
build a new steel works. I met the firm's engineer at Heathrow and
one of his items of baggage was a crate of a dozen empty bottles! He
was familiar with Algeria and knew that you could only get bottled
water, the only sort safe to drink, if you had empties to trade in.
When we got to Algiers he picked up the hire car and we
immediately drove east out of the city, as there were civil
disturbances there at the time. We were booked into a hotel perhaps
150 km away, just about the worst place calling itself a hotel I have
ever had to stay in. There was a dribble of brown water available
from the cold tap, but no hot water. It was also the period of
Ramadan, when Muslims are forbidden to eat or drink during
daylight hours, but they were prepared to serve foreigners with food
rather late in the day, when they were allowed to prepare food that
they themselves would only eat after dark. Meals were pretty awful!

Next day we were on the road just after some sort of breakfast
to inspect the site. First job was to find a shop open to stock up

with water and see what we could buy for lunch. A couple of packets of biscuits and some bread rolls would have to do. There had been a serious flood a week or so earlier and the section of flood plain which it was proposed to build up to a higher level had already had on it a deposit of gravel about 50 cm deep, because there the river had come out of a section confined between high rocky banks into a more open section where the flow spread over the flood plains and slowed down. It could therefore not carry the heavy load of gravel and sand that was within its capacity within the gorge-like section upstream. Local citrus growers were very upset because they believed their plantations would not survive having their roots so far below the new ground surface. This was an example of how river systems are shaped by the rare event. At lunch-time we pulled into a remote lay-by to have a drink and have a bite to eat, watching out in case there were any passers-by who might be Muslim fanatics. We were well into our simple repast when a group of hooded women walked along the road and we were quickly into the car and away before they might raise the alarm about foreigners not observing Ramadan! The trip back to Algiers for the flight home was along the Corniche, a road cut into the face of the cliff that forms that part of the coast of north Africa, quite a hair-raising trip because in places there were bends that a car could get round at full lock but the small lorries on the route could not, so they had to joggle back and forth a bit to negotiate the worst corners. Meeting any cars on a bend, or somewhat larger vehicles anywhere else, was indeed a bit frightening.

Binnie & Partners had a major job in Cairo, designing an entirely new sewerage system to replace the existing one which was overloaded and virtually full of sand, so that there was raw sewage lying on the streets in several sections of the city. My job there was to advise on design methods that would ensure that the large replacement collector sewers would be able to convey the silt load that would inevitably get into them, either through wind-blown sand washed in from the streets or from the domestic properties, where it was common to use sand to scour dishes. Although there had been research relevant to the power industry on the transport of high concentrations of solids in pipelines, and some research on the gradient a pipe would need to be entirely free of any deposit, these were not applicable to the case of the Cairo sewers. The former was for pipes flowing full, the latter imposed a very severe condition where, as in our case, all the flow had to travel long distances to pumping stations, implying heavier pumping costs. It seemed better therefore to modify the Ackers and White function for sediment transport in open channels to suit the non-rectangular section of flow in a part-full drain. It seemed reasonable to allow a small amount of deposition to give, say, a deposit with a bed width of ten per cent of the tunnel diameter which then defined the active width of sediment bed. A little algebraic manipulation then gave the governing equations, which did not fit too badly when the results of later research were available for comparison.

Egypt was interesting in so many ways, the pyramids at Gaza for example where I was able to go to see a *son et lumière* show one evening. On our weekend off, I was taken south to see the even older pyramids at Saqqara and we went into the underground tomb where there are huge figures of bulls in basalt each weighing many tonnes, the manoeuvring of which in the confined space of the tunnel system must have been a most difficult task for the ancient workmen. Some never got to their intended position as a result. One of the minor things that stuck in my mind from that trip into the desert is drinking the warmest beer I have ever tasted with our picnic lunch!

Iran and Iraq are much in the news now, and I visited both in better days. Mind you, even then one was well aware of the different lifestyles in these strictly Muslim countries from those in the Western world. One example of the dangers of being in the wrong place at the wrong time was brought home to me on a visit to Iraq, when a small group from Binnies flew to Baghdad from where we drove north to Kirkuk, which was to be the base for our study into an irrigation project in the valley of the Lesser Zab River. On arrival at the only hotel where foreigners could stay, we could not help reading the two large notices in the hotel forecourt – or at least the one on the right, the one on the left being the Arabic version. Each notice was about 2 m square, and began by praising the 'glorious' (sic) Batthist revolution during which all the Jews had been murdered together with their American capitalist friends. (I wonder if President Bush once went there and hence his

hatred of the Saddam Hussein regime?) We walked outside the
hotel after work one day and came face-to-face with the proud
character of the Kurds. There were two tall men in camouflage
gear coming towards us, with Kalashnikovs over their shoulders.
They were not going to step aside for foreigners to get past: no, it
was clearly up to us to get out of their way by stepping into the
road. So Iraq was not a country to feel welcome in, but we had
our job to do which luckily meant spending most of our time out
in the rather barren landscape. In fact, when that revolution took
place, Binnies had staff who escaped northwards to Turkey, Iraq's
neighbour, but our visit went smoothly.

There had been some very ancient irrigation schemes in this
area, including an attempt from perhaps 500 years ago to transfer
water from one valley to the next, never finished, but remnants
could still be seen. Our job was to pick up where a more recent
scheme had been suspended, to review and upgrade the scheme,
to provide a large lined channel, so we had to look at the probable
route and decide what needed to be done in terms of survey work
and detailed engineering design. The hydraulics of the system were
significant of course. Whilst in the northern region of the country
we also went to see the Dokan Dam, an arch dam over 100 m high
completed around 1960, higher up the Lesser Zab river. All
seemed to be in good order.

Iran is a significantly bigger country, about 800 miles from north
to south and width much the same. The capital, Tehran, is only
seventy miles from the Caspian Sea, from which it is separated by
the Elbubi Mountains. In fact Iran is a very mountainous country,
a lot being over 2,000 m above sea level. One visit there was to
advise on the location for a desalination plant near Bandar Abbas
on the Strait of Hormuz, which would require intakes and outfalls
into the sea. This was to provide drinking water to proposed
developments in the area. We flew to Shiraz, a city holy to
Muslims with a fine mosque, tiled basically in blue as is the local
Persian style. There we picked up a small plane, perhaps an
eight-seater, a Britain Norman Islander, planes which were made
in the Isle of Wight though later versions were built in Brazil,
including the Trilander with three engines rather than two.
Extraordinarily we had an air-hostess with us – but not to serve

our needs, she was the pilot's girlfriend. It was about 300 or 400 miles to Bandar Abbas, which then did not have an airfield so we landed on a strip of new dual carriageway road, part of the new development but not yet in use! We did our site inspection and set off on the return journey over very mountainous terrain, with a good deal of air turbulence. Obviously the pilot could see what was ahead; he could fly round the storm clouds but not over them. He must have judged a storm to be on the way, so suddenly we were descending and he landed on a military airstrip, just as the wave of wind-blown sand and rain arrived, in the nick of time before visibility closed in. There was just one bewildered soldier keeping guard so the pilot borrowed his bike and set off to the local town or village to find a vehicle that would take us there until the storm blew over and we could resume the flight. Was this an emergency landing – or did he always have in mind an intention to stop because he seemed to have friends there?

Another visit to Iran was to inspect the Reza Shah Kebir dam – though by now it has a different name to disassociate it from the Shah's reign. It is a high mass-concrete dam with a power station below it and its spillway going down its steep downstream slope into a flip bucket to throw the discharge clear of the dam. This had been an American design, and I think they also supervised its construction. Severe cavitation damage had occurred on this downstream face of the dam even though it had never carried anything like its design discharge (Figs. 10 & 11). Cavitation causes pitting of the concrete, which can be progressive and there were many places where it had gone at least 30 cm deep. It is the technical name for a phenomenon that arises in fluids when the pressure is below the vapour pressure of the fluid, so that in effect it cannot hold itself together; small pockets of water vapour form: it boils! These bubbles then collapse where the pressure rises slightly, and it is the collapse that generates huge localised pressures which pick away the concrete surface. Geoffrey Binnie was one of the party inspecting the dam, near his eighties by then but still quite fit and with a very active mind. He seemed to scramble up the steep slope better than I did – I haven't got a good head for heights and this was a high dam. In fact, that was the root of the problem. When there is flow over a dam crest, this accelerates as it goes down the slope, converting potential energy to kinetic energy, opposed only by surface friction. At the time the dam was designed, it was known that velocities over 30 m/s could cause damage, and although the designers had had a model built and studied, they obviously did not appreciate the risk they were taking with velocities of up to 40 m/s. Research in the US had shown that a very high standard of surface finish and straightness was required to avoid cavitation at such high velocities, but the nature of civil engineering at the time was that larger and higher dams were being built, meaning higher velocities too. We had a long steel straight edge with us and soon established that the standard of finish of the spillway surface was nothing like good enough. The damage could be patched with stronger material – but almost certainly would fail again. Nowadays, spillways of this type and this height are always provided with a means of getting air into the bottom layers of the flow, to provide a cushion to these cavitation pressures. There was enough information in the literature about recent research into cavitation, including an index to its likelihood, for it to be possible to link together all the governing equations: the basic hydraulics, boundary layer development, air entrainment from the surface by turbulence and the cavitation index, in a computational model. I believe my team at Binnies was the first to develop and utilise such a model and it did indeed confirm that this Iranian dam was liable to cavitation damage exactly where it had occurred. The problem there was that the designers were not up-to-date with the latest research.

On our return to Tehran, our local contact there asked if we would like to see the crown jewels. Yes please, if it is possible! I suppose we were a little surprised to discover that they were intact and available to view. We were taken to a back street and entered a rather run-down building, with a janitor sat near the door. He was a friend of our contact, and after taking the glass of tea that is the usual courtesy in the Middle East, the big steel door of the secure room was opened and in we went. There was the Peacock Throne, there were the crowns and other jewelled gold items, neatly arranged in glass display cases, an absolute Aladdin's Cave of riches. I wonder if it is still there and still intact!

CHAPTER 10

The Indian sub-continent

BINNIE & PARTNERS HAD LONG HAD significant involvement in Pakistan, for example having been principal engineers in the design and construction of the Mangla Dam on the Jhelum River, a major tributary of the Indus. This was already complete when I joined the firm, and was one of the largest dams in the world at the time of its construction. In Pakistan, dams are multi-functional in that they store water in the wet season, for release for irrigation in the dry season, and also generate power as this water is released. The fact that the rivers in that area are also fed by snow melt means that water is available over the whole year provided its release is properly managed. The Pakistanis like the Indians have many very experienced and competent civil engineers so that new projects are usually conceived as joint efforts. Overseas consultants are brought in to gain from an even wider range of experience and perhaps to benefit from their understanding of latest methods. As the B & P hydraulics consultant I therefore made many trips there in connection with new hydropower and irrigation projects. In fact, there are so many results from the past involvement of British engineers there, particularly irrigation canal systems built in the time of empire, before partition took place. These were still working well but some of the irrigation canals were in need of maintenance, and the river barrages that fed them needed refurbishment with new gates, for example. Partition of the Indian sub-continent meant that the needs for irrigation water in the two new countries would require some new configuration of reservoirs and main canals. Mangla was one of these, as was Tarbela on the main stem of the Indus, built a little later with American consultants taking a major role.

During my period of involvement there were two hydropower projects on the Indus downstream of Tarbela Dam that exercised our minds and engineering skill. The first of these was to be a typical high dam impounding a reservoir, and the site was

downstream of the Attock Gorge. It was already clear that although
Tarbela had been in operation for just about ten years, there were
serious problems looming about sedimentation within the reservoir, which would not only reduce its storage capacity, but would
also imply that more abrasive sand would pass through the turbines
to give severe maintenance problems in the future. This was not
our problem, but when Tarbela ceased to be an effective sediment
trap, all that material would pass downstream to the next reservoir
in the system i.e. the Attock Gorge proposal. This sediment was a
major problem to be tackled. It was notoriously difficult, if not
impossible, to make a physical model of the proposed reservoir in
which sedimentation and scour of a wide range of material sizes
would be correctly simulated, representing perhaps 100 years. Valid
sediment modelling could be carried out in certain restricted
circumstances but not in this case. Numerical modelling was much
more appropriate and the team I had built up had the necessary
skill in this. There was a commercially available model developed
in America which represented flow and sediment in a river system
carrying sediment. Would this suffice?

The model had been developed primarily to assess dredging
requirements in a river like the Mississippi with a navigation
channel of known width. To set up the model it was obviously
necessary to have full survey data to represent the geometry as a
sequence of cross-sections, also on the flow regime expected
through dry season to flood season year by year with some years
wetter than others as in nature. Then there was a need to know
how much sediment would come in from upstream. Although
there was a choice of sediment transport equations already in the
model, the first modification was to incorporate the Ackers and
White equation, the one we had confidence in. Initial tests soon
showed the major flaw in this model. By assuming that all sediment
movement was restricted to uniform bed level changes over a fixed
width, after a few years of sediment deposition this fixed width had
risen above water level, perhaps tens of metres, with the water
flowing either side of it, yet sediment was still assumed to move
across this strip above water level! A radical rethink was needed and
the solution was so simple that it is surprising it was not already
part of the model. It was to assume that sediment could be

deposited or scoured across the full width of the flow at any time-step in the model, proportional to the local water depth. It was a simple bit of computer programming to incorporate the necessary algorithm in the model, so updating the cross-sections at each time-step. We also accelerated the sediment movement by applying a multiplier so that each time-step in the model represented a longer period in the prototype. This reduced the running time of the model. This testing, upgrading and development of the computer program took place in the London office but the actual operation of the model would be carried out in the Lahore office, jointly with our Pakistan counterparts. In those days, the early 80s, our computers were mainly desktop Hewlett Packards, quite powerful for their day, which had to be fed with the programs and data on magnetic tape, and the output came on line printers. The same machines were available in the Lahore office but in terms of power and convenience these machines could not compare with today's PCs.

When this numerical model was applied to the Indus, the two reservoirs, the existing Tarbela one and the proposed Attock one, were ultimately linked together but the first step was to use what was known about Tarbela to check the working of the model. The volume of sediment entering was reasonably well-known from the survey information about how much had been deposited in the ten years or so of operation, and there was information from upstream from data collected there and knowledge of what the upstream bed was composed of. Putting all this together and running the model showed remarkably good agreement, both in the location down the length of the reservoir and the general form of cross-section. It was common knowledge that a sedimented reservoir has a very distinctive shape of cross-section, deep in the middle and with convex higher flanks, caused by deposition across the full width with scour only in the remnant channels when drawn down to a lower level. The new model produced this shape very convincingly. It was not surprising that the total volume of sedimentation matched because it was so closely related to what came in. The model represented some ten sizes of sediment (the Ackers and White equation had been developed to cover a wide mixture of sizes) as well as fine suspended material. We could now proceed to look at the long term. How would the sediment passing through Tarbela in the future affect the Attock scheme downstream? The model included the probable operating rules during a typical year, impoundment up to top retention level, release during the irrigation season down to a low level, and also a possible sluicing period when, with there still being water available the level would be drawn down by opening low level sluice gates to result in fairly fast shallow flows to scour the previous season's deposition. I believe this was a most successful operation of a numerical model to represent reservoir sedimentation and B & P were ahead of the world in this sort of study at that time.

Meanwhile, the Pakistan Water and Power Department had set up a huge physical model which we visited. This was perhaps 200 m long but even so was to a small scale as it covered perhaps 100 km of river valley. I don't think it could ever have been claimed to be a success, because the problems of simulating a wide range of sediment sizes at such a small geometric scale was not really resolved, and probably never will be.

Lahore was not a bad place to visit and spend some time in, which I did on many occasions at different phases of projects. There is an old part of the city, with narrow streets and water supply pipes snaking down many of the streets. A bit messy but it seemed safe enough when I went there with a colleague, John Pitt, to find a gallery he had read about with miniature paintings of scenes of Indian life some time back. We found it and the old man in charge – I think the owner-occupier of the small old house – was very courteous and pleased to show us his collection at no cost. The newer areas include a park and zoo, and a cricket field used for test matches. There are some modern international hotels, and the American Embassy ran a social club. There are (or at least were then) several restaurants including a French-style example with reasonably good food. There is a shopping mall, with a department store with a wide range of goods on sale. B & P had a rest-house with about six bedrooms for short-term guests and also any longer-term visiting engineers, so it was not necessary to use those hotels. The rest-house gave a good deal of freedom. The food was prepared by a male chef, and there were several other male servants. We could have a beer or gin and tonic if we wished, because long-term stayers could register as foreign visitors unable to survive without alcohol! We had a fairly permanent senior engineer there, Clive Baker, who took this upon himself and every few months he would make the trip in the firm's Land Rover to Islamabad where there was an alcohol store behind one of the hotels, approved by government to provide the ration. Having a cool beer when the temperature outside was about 40°C was a treat, especially after walking perhaps two miles from the office at the end of the day's work at about 5 p.m. The fridge was an important item therefore but power supplies were rather erratic, with low voltage sometimes and occasional transformer faults which reduced voltage to not much more than half the usual. Shortage of power was one of the reasons for the hydro schemes in mind, of course, getting worse all the time as more people expected and could afford air conditioning. We had to respect local customs and once, when John Pitt came back from playing squash, he sat in his shorts on the veranda and was seen by a neighbour who complained about him being improperly dressed, so he had to be careful to change into slacks in future.

There is a large mosque of red brick, and Westerners could visit it if they wished. It had high walls enclosing a courtyard, and one left one's shoes on a rack just outside to walk across to the mosque itself. There was a jute carpet across, and this was continually being doused with water to keep it cool. If you wandered off it in the hot season you would burn the soles of your feet! Outside the city are the Shalimar Gardens, dating from Mogul times. These are ornamental layouts with small canals crossing courtyards and grass areas but I never saw any water flowing in these channels. There are several buildings in the very ornate style of those times. So Lahore was a pleasant base for our work, from where we could go, usually on day trips, for visits to the site or the hydraulics research establishments some 100 km away perhaps. On one trip out of town, we had a strange experience. Our vehicle had United Nations' markings, and going through a fairly built-up area there was a group of men on the left of a crossroads. As we approached, one ran into the path of the vehicle which any sighted person would have seen. Our driver swerved and braked – he was going fairly carefully anyway – and this man actually ran into the front wing as we drew to a halt. What was going on? It became clear that this blind person's keeper (if that is the word) had told him when to run into the road; he had seen the UN insignia and reckoned we were foreigners with cash to meet any demand for compensation. The runner was not hurt of course (or if he was it was because he bumped the stationary vehicle). Our driver quickly sized up the situation and told us to stay in the car – John Pitt was all for getting out to check he was all right – and quickly drove away in case the little crowd that were there to watch the action turned nasty. This just goes to show what hazards there are for the unwary and that the locals should be trusted to show good judgement in these things.

Although there was always this need to respect local customs, the form of Islam in those days was not extreme and at times it could surprise you in its exceptions. Once we were invited to visit the home of a Pakistani who was an acquaintance of Clive Baker, and when we arrived at the compound where he lived, and rang the doorbell his wife came and told us that her husband was not yet home, yet against all the rules of strict Islam she invited us in to the lounge to have a beer and await her husband while she chatted freely to us. There was no question of the all-embracing clothing that the wives of rigid Muslim men are required to wear, and no hesitation at being alone with men she did not know. Clearly she was well educated – I seem to recollect a professional e.g. a doctor – and the family were thus quite relaxed. The walled compound included several family homes and so they had their own local private environment and security.

In the event, WAPDA, the Water and Power Department Authority, decided not to proceed with the Attock scheme. The impoundment and top flood level implied would have raised the flood levels in a tributary valley where the city of Nowshera flanked the river, and although our model had predicted these levels and shown that it would have been easy to raise banks to protect the city, probably the fact that they were in different administrative districts increased the difficulty of getting agreement. Instead, a quite different type of hydropower scheme came to be considered downstream of Tarbela. This would be a barrage not far downstream from the foot of the Tarbela spillways, at

Ghazi, feeding into a canal on the left bank which would continue many kilometres across the plateau until it would reach a hydro-plant at Barotha where the head available above the river downstream of Attock would provide the generating head for a power station down near river level. This presented many challenging problems: a flood flow equalling that provided for at Tarbela; a gated structure to pass this flow without causing erosion downstream; a gated intake that would exclude as much sediment as possible from the power canal; the design of this canal to pass the required flow and to keep clear of sediment (Fig. 12); the forebay leading to the pipes feeding the turbines; emergency spill arrangements to cover the situation where the power station had to shut down suddenly whilst there was so much still coming to it in the canal (Fig. 13); the energy dissipation structures dropping this flow down to enter the Indus again; and the many drainage systems that would be intersected and would have to cross over or under the power canal. Although I was involved with all the hydraulic problems thrown up as this scheme was developed, including the design of the largest siphon spillway in the world as the first element of the emergency spillway from the power canal to cover sudden shutdown, the seven model studies that became necessary were in fact the concern of civil engineer son John, who was the firm's chief hydraulic engineer much involved with the project after I retired. He also spent many months in Pakistan as the detailed design was developed, and later during its construction and commissioning.

Whilst I was still involved, we set up a two-dimensional computational model of the head pond to determine how it might stabilise after any sediment released from Tarbela entered it. Also a mobile bed physical model was designed of the whole of the headworks to examine how well the design would work in minimising the sediment entering the power canal, and all aspects of the performance of the main spillway and an ancillary spillway for extreme events on the right bank. This model would use a light weight sediment and was designed by comparing the transport rate in the model with that in the prototype at similar depths and velocities according to the Ackers and White formula, whilst retaining the necessary principle of Froudian similarity i.e. getting gravitational and inertial effects to the same scale. This model was quite successful, and it was feasible because it was not necessary to represent in the model channel friction as the model was short. In fact when this model and models of the barrage area and the emergency spillway from the power channel were carried out, it was son John who supervised them during the period when he was resident in Lahore. He was also there during much of the construction period of this huge project and again during its commissioning. This gave a rare opportunity to see all those features that had been modelled in action, the emergency release from the power canal carrying over 1,600 cubic metres of water per second, very many times more than has ever been carried by any other siphon system. (To imagine this flow, think of some three times the highest flood ever recorded in the Thames). The flood spillway at Plover Cove in Hong Kong of almost the same overall capacity had been designed for quite extreme events and had never been subjected to anything approaching its full load. The Barotha siphons were a bit slow to prime fully but this was probably a result of a very poor standard of construction. Some even lacked the step that is essential to initiate priming. John was not involved in the supervision of the Chinese contractors, which was the responsibility of the American resident engineer from Harza, the US firm that were co-consultants on the scheme.

There was naturally always a bit of tension in the background when working in Pakistan because the country had many engineers who were experts in the same fields that we foreign consultants were giving advice on. The client, WAPDA, had some within their organisation whose opinion they would seek about the work we had done. Their hydraulics man, for whom I had great respect, was not too happy about the model scales chosen for the mobile bed model of the headworks at Gazi Barotha and so, after its completion, recommended a repeat of the work, based on his reworking of the sediment transport equations, which gave a slightly different vertical scale, slightly modified sediment size and feed rates, and a modified slope exaggeration (although simulation of the natural gradient was not really significant). My experience of physical modelling was that a pretty good model would give reliable results even if it was not a perfect simulation, and this proved to be the case here. It does not matter which scales were more accurate, the results in practice were virtually indistinguishable but honour was served!

The major part of the Indian sub-continent is of course India itself, and the only involvement there on Binnies' behalf was concerned with the new water supply to Bombay, Mumbai as it is now called. My only recollection of that project was seeing the large pipes, probably about 2 m in diameter being used as temporary homes as they awaited installation. India is a country of contrasts with many very poor people, beggars trying to eke out a living, and the caste system that separates people rigidly into categories of work they might do, with the horrid connotation of the 'untouchables'. Two significant visits were to New Delhi for an IAHR congress, the biennial main technical meeting of the International Association for Hydraulic Research, which meets in different countries in some sort of rotation and which I attended whenever I could. The other was to Poona, to give a series of lectures at the Research Institute there. One only had to mention that one had worked at HRS Wallingford in the days of Sir Claude Inglis for the engineers and scientists there to express their delight at meeting someone who not only knew him but had worked with him. It was Poona where Sir Claude set up the Indian Irrigation Research Institute with colleagues Ted Crump (also an HRS colleague) and Gerald Lacey, who developed a theory – or was it really a set of design functions that expressed his understanding? – of river and canal regime. I have never been quite sure whether there was a deep theory hidden there if only someone could expose it! The work of those expats was still greatly appreciated and I never found any resentment at all about Britain's imperial past insofar as it affected India or Pakistan.

My stay in Poona was interesting because I was housed in a modest hotel in the old town for convenient access to the laboratory. There were local customs to observe, such as the wedding procession that passed under my window, with much trumpeting and clashing of cymbals, with the groom on a white horse in fine silk robes. There were very elaborate Hindu temples, and many little household shrines to the local gods; lots of pedestrians crowding the sidewalks; bicycles, handcart, *bejaks* (little

three-wheeler taxis based on a motor scooter, common in much
of the Far East as well). One felt quite safe when out walking on
one's own. The Western-style hotel was the Blue Mountains,
which had a bit of a reputation for cannabis smokers and the like,
and had been a sort of headquarters for the hippie set when they
were keen on an Indian sub-culture. I think the Beatles had used
it but it had reverted to just an ordinary hotel, which I could get
to by taxi if I felt like a change of food and location over the
weekend. One evening I was invited out to join one of the
laboratory staff and his family for an evening meal, which was a
very enjoyable experience and for which they had gone to a lot of
trouble to make me feel at home. The actual laboratory at Poona
had a wide scope and good facilities for both model studies and
more basic research. They had a good supply of water from a local
large canal, so could use large scales in their models. Many were
representations of rivers to study sediment movement, the performance of barrages and headworks, scour etc. There were also
larger models of spillways from dams than HRS had ever been able
to carry out at Wallingford, with its limited water supply.
Instrumentation was not quite so advanced but they were not far
behind the best European laboratories.

Being south of the crest of the Himalayas, it is appropriate to
include Nepal as part of the Indian sub-continent. The job there
was the preparation of an irrigation manual that would cover all
aspects of the design and operation of the whole range of systems
from the size of a field up to regional schemes. Of course larger
schemes would all have to be studied in detail but at least the
manual would indicate the features that would have to be
considered, e.g. the hydrology of the area, the river intake and
main canals. My role was to deal with the impacts of sediment on
the design and operation, whilst other specialists dealt with
different aspects. An office was set up in Kathmandu, with local
staff as well as expats, with support from ODA, the Overseas
Development Agency, and visitors were accommodated in a
nearby Western-style hotel.

Kathmandu was another place on the hippie trail, having a
reputation for the availability of drugs for the Westerners with that
sort of lifestyle in mind. They tended to be backpackers using

cheap overland transport and low cost accommodation. Nepal then was safe for foreigners although there were already signs of impending unrest in some areas where there were very left wing anti-government feelings. Kathmandu was an interesting place, with a temple area where several different religions were represented. One of these was the Tantric sect of Hinduism that featured elaborately decorated tall pyramids covered in highly coloured statuary of people, animals and gods, some of the humans being couples in erotic poses – but these were at the top or on the side without a good viewpoint so that the tourist would not see them! Many of these temples and memorials were gilded, and also there was a building nearby, around a courtyard where a young woman lived – or was incarcerated – only showing herself to the public by appearing on her balcony one or two days a year. This was some religious order and catching sight of the virgin was supposed to be a great thrill for believers. This building like many others was timber-framed and all its windows had elaborately carved tracery shutters (Fig. 14). One evening the lads from the office went out for the evening to have a meal at a restaurant with a reputation for a speciality main course. This meant taking a taxi to the other side of the virtually dry river Baghamati. Not a posh place and when this delicacy arrived I was able to say I had experienced this food before. It was a lamb stew, scouse, the Liverpudlian staple diet that led to the nickname 'scouser'. It is basically of Irish origin, the full name being lobscouse! There were in fact many eating establishments in the city because of its popularity with the lower end of the tourist market, and they did a good trade and provided a change from the unappetising Western-style beloved of many international hotels.

Nepal, which forms the southern slopes of the Himalayas, is geologically young, thus has steep slopes which are unstable and so landslides are common. Towards the southern border it is a bit flatter where it forms part of the Ganges valley. Rivers have a lot of melt water and run very steep until they approach this main valley. The preparation of the manual meant that we needed to inspect some typical irrigated areas, including some small scale ones so that we could make recommendations for sustainable systems i.e. those that would require minimum maintenance to keep working.

The other feature of Nepal that seems surprising to visitors is a general inaccessibility, and in fact there were no roads at all until about 1920, when the king was sent a gift of a motor car, which had to be separated into component parts so that it could be carried by porters to the palace in Kathmandu. Traditionally porters had to be used to move people about – or you just walked and climbed. During our visit no road existed along the length of the country to join its western border to the capital, though one was under construction with support from several European countries, the UK providing support through ODA for much of the highway, and a bridge over a sizeable river in the west being built by the Germans.

To help with our work we needed local knowledge of existing irrigation projects and the first field trip was within a day's excursion from Kathmandu. This illustrated very well the problems resulting from the development of the landscape. That area of Nepal, in the foothills of the Himalayas, is what we would regard as very mountainous, and having roads cut into steep hillsides, winding around the contours as they passed from one valley to the next, with quite steep gradients in places. These roads were frequently cut by landslides, which would either block the road with rocks falling down from above, or even whole hillsides tumbling down into the valley. We had to negotiate a number of these difficult places where the highway maintenance teams had just re-opened a section of road even though repairs were not complete and one still had to drive over a rough surface, or a narrow section. But the views of the Himalayan peaks forty or fifty miles away were stupendous, especially late in the day when the snow covered tops turned pink.

For the next field excursion, we needed to go towards the western end of the country. We flew to a small provincial airport at Pokhara so as to lessen the road journey. Although Nepal may look a small country on the map, we had between 400 and 500 km to cover, rather a lot for one day in that sort of terrain. Two four-wheel drive vehicles had gone ahead to await our arrival. You need two in remote areas in case of a breakdown. Then we drove to a place near a main road crossing from India where there were facilities and a modest hotel (Bhairawa) to spend the night before the next day's drive. Because it was near the crossing from India

12. The Ghazi Barotha power canal

13. The emergency spillway at Barotha under test. The siphons are at the top

14. *Typical houses in Kathmandu*

15. *Crossing the River Karnali by dugout canoe*

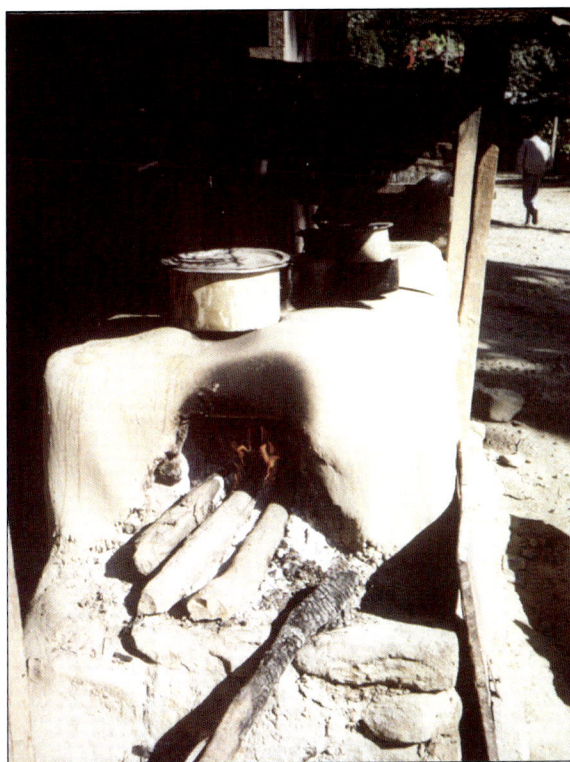

16. *The tea-shop stove in Nepal*

17. The Himalayas on the skyline

18. Siphon spillway at Plover Cove reservoir, Hong Kong

19. Dolosse *slope protection at High Island seaward dam, Hong Kong*

20. *Truck carrying a dolosse block from the casting yard*

21. *The Yellow River, 600 km from the sea*

22. *The power station at Sanmexia*

23. *Our locomotive for trip to Sanmexia*

24. *Underground farmhouse in the Loess area*

25. *Irrigation intake on Krueng Baro, near Banda Aceh*

26. *Dust cloud from volcanic eruption, Bandung*

27. The layer of dust outside the hotel

28. The weir controlling the irrigation intakes, River Gumbasa, Sulawesi

29. *A rice barn in Tana Taraja*

30. *Effigies of ancestors at burial place, Tana Taraja*

31. The River Ok Tedi, Papua New Guinea

32. Helicopter trip to collect samples from the river bed, Ok Tedi

33. Core samples, Ok Tedi

34. Secure Bay, Australia: location of proposed tidal power scheme

where many of Nepal's imports came from, the small town had a well-stocked general store where we could buy food for the next day, and the drive took us along the route of the new road being built towards the west. Part of this was through a forest where there were reputedly tigers and other animals. In fact there is some tourism to the region that claims to find a tiger or two but we saw nothing the whole day, no animals and hardly a bird either, no doubt sheltering from the heat and keeping well away from any signs of life. On this journey we were pretty well alone in our two vehicles, just passing one or two trucks going the opposite way. Some of our drive was on the foundations of the new road, but much was just a dirt road through this fairly sparse forest area. Our destination was the other side of the Karnali River, where there was a vehicle ferry – only when we got there it had broken down. It was a barge with ramps for access capable when working of taking say four vehicles like ours or a couple of trucks, and was operated by a chain across this fast flowing river. The only accommodation was on the other side, an old army barracks near the village on that side. What was to be done? Clearly the vehicles would have to be parked but the locals had a passenger ferry in operation, which was a dug-out canoe made from a single tree trunk, which would take about three passengers in addition to its crew of about four paddlers. The procedure was to tug this canoe upstream as far as possible before getting to a steep rocky bank, load it up then set off paddling like fury upstream but at a slight diagonal so that gradually it worked across to the other bank perhaps 100 m downstream (Fig. 15). The river was carrying quite a big flow from the melting snow and was very cold and somewhat frightening. Several trips and we were over with our minimum amount of luggage. Then it was a hike to the old barracks where army beds with straw mattresses awaited us. Simple food was available but first we refreshed ourselves with a drink from the local tea-house. This had a fire in a clay oven, with a pan of boiling water on the top. The tea was made in a pot and poured into glasses the rims of which the proprietor carefully wiped with the palm of his hand to make certain they were clean for these foreign visitors! (Fig. 16)

After a night's sleep we were up early. The ablution block was outside with a row of showers on a wall behind a screen, just simple sprays on a fixed head about 2.5 m high. The water was direct from the river, just above freezing, and it provided the coldest, quickest shower I have ever had. A freezing cold shave too.

One of our party had been out with a Nepalese colleague to see whether we could continue without our transport and they discovered there was a bus going the right way at about 9.30 and we got on, somewhat to the surprise of the half dozen or so local people who were using it, one man having a fighting cock with him and others with a few goods from the local merchant. The village was perhaps twenty simple dwellings and the shop which was also the tea shop on the T junction where the track to the ferry met another dirt road. We checked that the bus would return at about 4.30 p.m. from wherever its final destination was and we travelled perhaps ten miles to head up a river valley where there was some irrigated land and where we could see the scale of these local projects and how the local farmers used ad hoc procedures to divert flow from the river. Some of these were no more than a low temporary dam built from local pebbles and small boulders to raise water levels by perhaps 0.5 m as the flood receded, thereby feeding an earth channel that led onto and along the flood plain for several hundred metres. There was little point in suggesting anything more permanent in these steep catchments with fairly unstable rivers. Some schemes, however, fed permanent canals following a contour to a plain further downstream and here we could give advice on simple intakes that would restrict the ingress of gravel. Our route to some of these areas was along trails cut into the steep hillsides in rather beautiful country. On one of these we met a couple of boys, perhaps nine and eleven years old, each carrying some goods, the bigger boy having a bag of rice over his shoulder. Where had they come from in this relatively uninhabited area? We could not converse with them, and all we could do was offer some sweets as they continued their trek, presumably going home with what was the equivalent of the week's shopping for the family. We were back at the road where we got off the bus in time to catch its return journey for another night in our very sparse surroundings.

What one remembers from most trips into the wilds is that the local people are so welcoming, doing their best for foreigners who they know are used to a much higher standard of living than they can ever aspire to. Increasingly though, their first priority of civilisation is to own a television set, which you might see working in a mud house or even in a village square or tent as a communal facility.

Our journey back to Kathmandu was a very long day, starting of course with the dug-out canoe ferry crossing to pick up the vehicles (Fig. 17). We called at the Gandak Barrage on the River Gandak, which forms the boundary between Nepal and India. The Nepalese side of the barrage was very relaxed and we could take photographs but the Indian guards on the other side watched our every move and were clearly not too pleased about out presence. Had we been on their side our cameras would have been confiscated as they are so very sensitive over the security of their national resources. Do they fear terrorists from Pakistan making use of details of installations obtained from tourists' photographs?

Returning to Kathmandu, we were able to complete the drafting of the irrigation manual, including diagrams of typical practice and some simple theory. Being a civil engineer with experience from around the world throws up some surprising illustrations of how small the world really is, and I have had quite a few of these. One evening we were having dinner in a restaurant when someone who had seen me from the other side of the dining-room came across and enquired what I was doing there. He was in Nepal working on a water supply scheme. Neither of us knew the other was there. Just a chance encounter in a far away place!

CHAPTER 11

Hong Kong and China

BINNIES HAD BEEN INVOLVED WITH WATER supplies in Hong Kong for many years, including the Shek Pik reservoir in about 1960, and by the time I joined the firm, an office had been set up there to deal with an expanding workload. One was the Plover Cove reservoir, a scheme I was familiar with as the model studies for its siphon spillway had been carried out by my group at Howbery Park. It was probably contact with me for this project and also for earlier model studies such as the bellmouth spillway for Llyn Celyn Reservoir on the River Tryweryn in North Wales, that led to the firm inviting me to join them as hydraulics consultant. Plover Cove was a reservoir reclaimed from the sea, closing off an arm of the Tolo Channel by building a low dam one side of a small island and a spillway the other side. It had a modest catchment area so the design flood was 2,200 cumecs (equivalent to tonnes of water per second), requiring sixty-four separate siphon units, the largest siphon spillway built anywhere until it was matched by Ghazi Barotha in Pakistan. Hong Kong was the site of many siphon spillways (Fig. 18). One was added to the Shek Pik shaft spillway to allow it to carry its flood flow with a higher normal retention level in order to increase its storage volume. This had also been model studied, including the representation of potential vibration in the hood, under my supervision at HRS. Plover Cove reservoir was to be fed with water from a range of intakes on steep streams, and each of these also had its special hydraulic features that really were novel when they were built. The flow was taken down to deep collector tunnels by drop shafts, and each had a flood spillway using basically the same siphon design as at Plover Cove. These vortex drop shafts had also been designed using the equations developed by Ted Crump and were model tested at Wallingford.

Another major reservoir in Hong Kong was reclaimed from the sea. This is at High Island, and came somewhat later. Its spillway also has siphons, very similar to those at Shek Pik but with their entry designed to withstand the waves that might be generated in a typhoon. The main impact of typhoon waves would actually be those from the South China Sea on the seaward side of one of the two high dams that formed the reservoir. Cofferdams, temporary dams built so that the water could be pumped out from between them, were needed here because the dam foundations would be below sea level. In most schemes the cofferdams are later removed but here the seaward one provided the protection against the extreme wave conditions the project would have to be designed for (Fig. 19). The forecast wave climate was so severe that protection with rocks would not suffice and there were several forms of concrete block available for use in these circumstances. The one chosen was the *dolosse*, a South African development, the name coming from the word for a sheep's knuckle, as this was supposed to be similar in its shape. For the design wave height of 25 ft, 7.6 m, blocks weighing 6 tonnes would be required, placed randomly but very carefully on the seaward face of the coffer dam (Fig. 20). This dam was low enough to be overtopped by these waves so it was provided with a heavy concrete cap, and a different, much smaller, block protecting the back face. The model studies to confirm the design were carried out in a flume at the British Hovercraft Company in the Isle of Wight.

When designing for some rare occurrence, one might hope that it doesn't occur for very many years, but probability theory means that there is an equal chance any year from year one of its existence. Almost contemporaneous, was the Sines Port development on the coast of Portugal, designed for a similar wave climate to our scheme, with slightly heavier *dolosse* protection if I remember correctly. Their engineers were unlucky: it suffered a very big storm almost the first year it was completed – and it suffered serious failure with sections of the protection being stripped away. Could there have been something wrong with the strength of the blocks, their design or placement that could also apply to those at High Island? It led to a bit of worry, but the investigations of what went wrong at Sines neither fully resolved my worries nor pointed to a problem specific to Sines and not applicable to High Island. I was never convinced that those investigations came up with the true cause of the problem. There was a significant difference between the two installations however: High Island was in fairly shallow water so that the protection went down all the way to the rock seabed; the Sines breakwater was in deep water so the protection rested at its lower edge on a shelf built into the shape of the main embankment. I thought it quite possible that the failure there was really due to a soil mechanics problem that caused this shelf to shear off in places and so let the protection it was supporting slump down. Many of the dislodged blocks were down at seabed level. The water pressures within the embankment under severe wave attack would have fluctuated wildly. The impact of waves can produce very high pressures as anyone knows who watches a wave breaking against a sea wall and producing high plumes of spray. We shall never know, but the High Island protection is apparently safe and sound after some thirty years!

There were regular calls for me to visit Hong Kong over the years, sometimes as a stop on a journey primarily to some other place in the Far East, so I was going there two or three times each year, and knew the city well. That was before the days of the cross-harbour tunnels or underground railway connecting the New Territories to the island itself, so there were many trips across Star Ferry, and rides up to the Peak at the weekend. It also became very much a family concern, because John was seconded there by the firm – he had by then joined B & P – working on a comprehensive study of slope stability where shelves had been excavated into the steep hillsides to form platforms for building, and there had been some lethal landslides as a result of tropical rainfall. Whilst he and his wife Lizzie were there, two of our grandsons were born, Thomas in 1978 and Robin in 1981. There was an amazing coincidence when they went to Hong Kong. Just before they were going they found out that a couple they were very friendly with were also going at virtually the same time. That made three because another couple of close friends were already there. The male friends had been at Imperial College doing the same course as John; the wives had been nurses with Lizzie! Then not long after they left to return to the UK, with John then being in the firm's Chester office, our daughter Sheila went there. She was a systems analyst in an American bank that found it convenient to have the British based team as support for the computer services e.g. in

Singapore, Korea and the Philippines. Margaret was able to accompany me on some of these trips, and we took several journeys out of Hong Kong, both with John and family and with Sheila. I was once able to meet up with Sheila in Singapore for a weekend, when she brought along a young lady friend (whose name was Gaynor Dowd but she was certainly no dowd!). Going for a swim on Samosar Island with these two bikini-clad lovelies must have made some watchers wonder whether I was some sort of sugar daddy to these comely lasses, rather than the actual daddy of one of them!

Another project was the development of Discovery Bay on Lantau Island, which encompassed many different aspects of civil engineering: its water supply, involving a dam to form a reservoir with its essential flood spillway, water treatment plant, sewerage system and road works. Basically it began as a golf resort, with hotels as well as the course itself, but developed into a new beach resort. It included a promenade and, as there was no natural beach there, one had to be created. This was an unusual project and entailed importing dredged sand to form an arc between the two headlands on either side. As Hong Kong can have very high rainfall, we relied on our hydrologist colleagues to provide an estimate of the maximum flow for designing the spillway and drains. The spillway from the dam has to be designed for the probable maximum flood (PMF), which is the flow resulting from runoff of the highest conceivable rainfall ever possible. All spillways from dams have to be designed to discharge safely such an extreme flood flow. Otherwise a calamity would result; if the dam itself were overtopped, the whole structure could be washed away, releasing the water it stored in a giant wave.

The design of the beach was straightforward in a sense, as the likely wave direction was fairly obvious and it was a sheltered area making big storm waves very unlikely. If there was a concern, it was whether the gentle wave climate would keep the beach clean. Discovery Bay became an attractive venue for day-trippers, with a ferry from Hong Kong itself, and I was able some years later to visit it and see the development for myself.

One of the last projects I was involved with in Hong Kong was the major scheme for a new sewerage system, involving deep collector tunnels, many under water, primary and secondary treatment works, and then a long submarine outfall to a series of outlet pipes so diffusing the effluent over a large area of the sea and diluting it as the buoyant plume rose, mixing with the salt water around. My involvement was modest in that I gave advice on the self-cleansing velocities needed and on the spacing and design of the outlets, the mixing and dispersion in the tidal flow and the velocities needed to purge the system of sea water that could fill the pipes before they were operational and any period of shut down or very low flow. This was an important topic of research at that time, with work being done at BHRA, the British Hydromechanics Research Association, and later taken up with specific reference to the Hong Kong scheme by Joseph Lee, by then Professor of Hydraulics at Hong Kong University, who did his MSc at Imperial College when I was Visiting Professor there, and whose career I have followed with interest. Over the thirty-five years since I first went to HK, it has reverted to China, becoming a special zone within that country, and has developed even further. This of course was the reason for so much effort going into water supplies when they had to rely almost entirely on indigenous sources. Now, as an integral part of China, there is no political restriction on how much they might get from the mainland.

Much of the postwar development has been in the form of new townships, and during my time the drainage from the low lying land that was to be utilised for one of these schemes was quite a taxing problem in terms of its hydraulics, because it required a channel for the lower flows flanked by berms that would come into play in times of flood, a compound trapezoidal channel, the design of which was then very uncertain. Should we treat them as one channel, or treat the deep central section separately from the shallower flanks and then add them together? This later became a topic for research at my old workplace, which after privatisation became HR Wallingford, and as a result of my involvement with that research as a consultant to the lab I could now resolve that problem better, although how to treat naturally-occurring me-andered compound channels remains very uncertain!

I was always happy to go to HK. A few times Margaret came with me, sometimes I stayed with John and family, sometimes with

Sheila who had a flat in a new block facing the island, where land had been reclaimed and a new promenade built. There were so many places to visit, frequently using the good ferry services to the outlying islands. Lamma was a good place to go for a hike as one could get a ferry to one side of the island, cross over its high ground, to get the ferry back from the other side, perhaps having a meal at one of the waterside restaurants. John and family had a flat in a multi-storey block on the mid-levels below the Peak, within walking distance of the centre of the city and Star Ferry, but a steep climb back if you chose not to get a taxi, up streets with stepped footpaths that have now been fitted with escalators. There was a wide range of eating places, with any Western-style if you were so minded but excellent Chinese ones. It was quite usual to take Sunday brunch in one of these, *dim sum*, where you chose items from dishes being brought round on trolleys with steamers to keep them hot. Some were floating restaurants, in that they were on rather smart junks, but some were very modest such as in the market, where one floor was entirely taken up by separate eating places. You could eat cheaply at these, and in our experience safely. One such lowly eatery provided lunch for a party, including Margaret and my colleague David Ruxton, when we took a hired motor launch out to sea via the Tolo Channel to see the Plover Cove spillway siphons from the outlet side and to look at the proposed High Island seawater dam site. Just off the coast was a small island – or was it just a shack built on stilts? – which obviously specialised in seafood. The fish, crabs, lobsters etc. were kept in tanks at sea level so the procedure was to choose your fish for the main course and sit and have a beer while you waited, perhaps with some spring rolls as starter. This was a quite filthy place. David commented that he could never bring his wife to a place like this but we were assured the food was good and safe. I was not particularly adept with chopsticks and dropped one. 'Don't pick it up! The floor's filthy. We'll get you a clean one' was the cry. And the fish was quite superb with no after effects.

On a later trip, Margaret, John, Lizzie and young Thomas went with other Binnies' families to a beach adjacent to the High Island seaward cofferdam so we saw the *dolosse*-protected slope after it was completed. It was a brilliantly sunny day to spend at the seaside, but my sunburn protection was defective. My legs were seriously burned, and the next evening I was due to fly home. I could hardly walk and spent the day trying my legs out whilst wearing only my pyjamas! I was just fit enough to make the journey to the airport and so home! I recollect that among the several Binnie families on the beach that day were the Ludfords. I had known Peter in the London office. I believe their parents were also there on a visit and much later in life when Margaret and I joined the local U3A, the University of the Third Age, a lady, hearing our name, came and asked if we were any relation to John and Lizzie Ackers whom she had met when her son Peter had worked with John in Hong Kong. We confirmed that indeed John was our son, but this contact would never have been made had we not had a somewhat unusual name.

There was recently another strange coincidence that went back to those trips to Hong Kong. One of my contacts there, Kishan da Silva, now lives in the UK following his retirement, and he has a vintage Rolls Royce. Our local U3A had a long weekend break last year based in Maidstone. On the Sunday morning, Margaret and I joined a visit to a nearby country park, but our neighbours, Peter Dunne and his wife, decided to stay at the hotel. A local veteran car club was starting a rally that morning and Peter went to look at the assembled cars just as I would have done had I been there. He got into conversation with the owner of one, discovered he was a retired civil engineer, mentioned he had a neighbour who was also a retired civil engineer, was asked his name, whereupon Mr da Silva said he knew me and gave him his card! Of course by the time the main party returned to the hotel, the cars had left, but this is one of several examples during my career where there have been similar coincidences showing that it really is not such a big world after all – at least if you are a civil engineer!

My first experience with chop sticks was in Singapore, where Binnies had an office, and there was a local partner. On my first visit there a dinner was arranged in a private room in a restaurant, with the wife of the partner and those of the other senior staff there. I had charming Chinese ladies each side of me who insisted that I learned to use chopsticks there and then, with their tuition and help as necessary. I did manage the meal and in so many later visits to the Far East some ability with chopsticks was necessary to avoid starvation! The worst ones I came across were in Korea in the office canteen of Hyundai. They were short stainless steel ones, not very efficient for eating rice or noodle soup, but as Korean food is at the other end of the scale of delicacy to Chinese perhaps it was not too much of a hardship!

Having mentioned Korea, this seems to be the place to describe my only visit there. This was in connection with a review of their hydrological data gathering, in other words what methods of measuring and recording the flows in their rivers they had, and the scope for improving them. We in Britain have always had excellent systems for collecting such data, sometimes using measuring weirs, sometimes level measurements correlated with a knowledge of how the flow rate varies with level at each site, occasionally using electromagnetic methods. The work at Wallingford on flow measurement meant that I could advise on one aspect, and there were hydrologists in the team to advise on other matters. Obviously we had to inspect many of the existing sites, which on the whole suffered from poor maintenance and some faulty design. We were at a hotel in Seoul at other times, and the capital was entering a phase of redevelopment, including a new art centre with a new opera house and concert hall. I was able to visit it during its opening festival, once to see The Royal Ballet from London in *Syncopations* and another night to hear Alicia da Rocha, a fine Spanish pianist. The hotel in Seoul was rather unusual in that on one side of the corridor were the Korean-style rooms and on the other, the Western style. The locals do not sleep in a bed but on a piece of carpet, rolled out on to the linoleum-covered floor, and there is a wooden rest with a curved top for their necks, not at all suitable for the less hardy Westerners, though even the Western bedrooms were rather stark.

The food is also on the whole not to our taste. One of their dishes has a name like *kim chi*, and consists of any vegetable matter they can collect from the fields and hedgerows which is put progressively into a big pot where it ferments. One evening we escaped to an out-of-town restaurant famed for a dish called *bulgodi* – or something like that. This was thin strips of beef cooked before us by young waitresses while we sat cross-legged around the central stove, on the floor, again covered in linoleum. There were various vegetables and rice to eat with it and it was quite good. The use of linoleum as the floor covering, rather than carpets, may stem from the fact that Korean houses are heated by under-floor ducts that convey the heat and smoke from peat fires lit outside, rather like the Roman hypocaust. The solid linoleum presumably seals the floor to keep out the carbon monoxide that would otherwise have a devastating effect.

One weekend we took a trip towards the east, intending to climb to the highest point in South Korea from where one would see the Sea of Japan. It was an amazing experience because it was a tourist attraction – and it really was a mountain to climb! The slopes were steep and many had vertical steel ladders up them, others were just steps cut in the rock. Whole families were there, including grandparents in their mountaineering gear and exceedingly fit, all very polite with much bowing when giving way to us foreigners. Was the view worth it? No! Clouds were coming over from the east and so we were in thick mist at the top.

My first trip to mainland China was in 1980 for an international symposium on sedimentation, perhaps the first international conference to be held in Beijing since it opened its borders to tourists after the terrible time under Mao Zedong's dictatorship. It was still not possible to fly to China except through Hong Kong, where one could get a Chinese Airlines flight to Beijing, which is far enough north to have very cold winters yet has a continental climate, so can be hot in summer. Also it is a very dry area, and even then one could have such still conditions that a haze of smog would build up. The conference included many experts on various aspects of sedimentation from all over the world so we had our first opportunity to meet the Chinese experts who had to deal with one of the largest countries, with big rivers. We made a major tour of the country at a time when most people in the provinces had never seen foreigners before – unless they had been unfortunate enough to remember the Japanese invaders many years earlier.

The outgoing journey was by train and coach, starting on an electric sleeper train which took us to Chengchou, where our dedicated sleeper coaches were shunted into a siding in readiness for the next overnight journey. These trains were quite pleasant, well-appointed and with a small kitchen at one end with an attendant whose job it was to provide us with cups of tea – black but weak – that the Chinese seem to drink all day. They were convenient too in that our Chinese professional colleagues were able to escape from their minders and come and talk to us. Most had good English, and it was notable that even the coach driver was studying English over his radio. These experts on hydraulics and rivers had escaped most of the privations during the days of Chairman Mao, as even the party officials recognised that their skills were essential to keep the country going in those dreadful times. Many millions living in the countryside starved to death whilst the 'upper classes' in the towns took their harvests, with ever increasing targets that they had to meet, and corrupt party officials kept the truth from the hierarchy. Despite this they had a good sense of humour.

From Chengchou the coaches took us to a section of the flood banks of the Yellow River which had been breached under the orders of Chiang Kai Shek in the civil war period to drown countless thousands in the plains beyond. The Yellow River carries the highest sediment load of any river in the world, brought down from the very erodible loess hills that form much of its catchment area. The river bed was perhaps 5 m above the surrounding plains, with flood banks perhaps 10 m high to keep it under control. There are a number of river systems in the world where sedimentation has raised bed levels and higher and higher flood banks are needed to contain them, in the Indian sub-continent and in parts of America for example. The Yellow River is so wide some 600 km from the sea that it can be difficult to see the other side! (Fig. 21) The flood banks are protected by groynes that have to be continually maintained so with that wide expanse of sand between you and the dry season channel it is more like a piece of coast! We also went to see one of the problems posed by such a braided river when the dominant channel moves away. From a prominent position, overlooking a stretch of river which was about 10 km wide, we saw perhaps a hundred of the local peasants digging by hand – there were no mechanical diggers in China, it seemed – a canal towards the dry season channel in an attempt to link it to the intake of an irrigation pumping station. The water was needed then as the planting season had come; without water there would be no germination. The pumps would raise the water to a second pumping station that would lift it further into a canal that ran at a higher level. We were also shown what a bad effect such sediment-laden water had on the pump impellers, wearing them out very quickly.

On our tour, we were being shown also the hydraulics laboratories of which there were many, under the auspices of universities or provincial irrigation departments, and one was at Loyang. This city boasted a large statue of Chairman Mao just in front of its town hall, which dominated the whole street scene. That was the next stop on our journey, the train then being diesel-hauled, and so on to Sanmexia, where there was a steam-hauled train waiting to take us down to the power station there (Fig. 22). Sanmexia is a gorge, the last place where a high dam could be sited before the river reached its flood plain beyond. This had been a Russian scheme from the days before the thousands of fellow communist advisers were chucked out! Rather surprisingly, their engineers had failed to take into account the huge sediment load, so that within a couple of years the reservoir had silted up, raising flood levels upstream to an unacceptable degree, and damaging the turbine blades. The Chinese engineers had had to step in and provide some low-level sluices and lower the normal retention level with a new flood spillway – no mean feat in an existing large dam. Our steam-hauled train ran on the branch line that had been used during the construction work, and was just one example of the many steam locomotives still in use in China, some with eight driving wheels and usually painted black with a red star at the front of the boiler (Fig. 23).

We were next taken to the areas in Shanxi province where serious gulley erosion was taking place. Some of the gulleys were huge, up to a kilometre wide at the top and a hundred metres or more deep. These were in the loess area, where, once the surface vegetation had gone, the basically sandy material below could not be prevented from being swept away by rainfall. It was this material that was providing the heavy sediment load in the Yellow River, of course. There was an action plan to reduce it which took the form of terracing, each terrace having a back slope and planted with apple trees. The area is so huge that it seemed a never-ending task. In those areas the farming practices were labour intensive, with threshing floors where a donkey would pull a log around to break the grain free and the wind would be used to remove the chaff. We also visited a village of underground dwellings. Imagine a quarry in the loess with vertical walls (the consolidated sand was almost like a soft sandstone which had quite a good strength and with vertical faces would not be eaten away by rain), with cave dwellings around its perimeter (Fig. 24). The residents were there in their Chairman Mao blue suits to greet us, but I reckoned that the party officials had had them all scrubbing and cleaning and hiding away their more meagre possessions before we arrived! A lunch stop in a town whose name I have forgotten was interesting. A line of about twenty enamelled wash bowls on metal stands was arranged at the front of the cafe forecourt, each with its block of soap and clean towel so that we could wash our hands under the gaze of the townsfolk assembled on the other side of the road to watch the foreigners. Food was simple but wholesome, and taken alfresco.

Then to Xian, from where we took a trip to some major hydraulic installations at Doujingyiang. There is an impressive barrage on the river Chung Ho, splitting this large river into two branches, one of which also feeds an irrigation channel while the other acts as the main route for major floods. The canal starts its journey via a channel cut into rock. This was one of the places where we saw the facilities and equipment used in China to measure river flows (using cableways supporting current meters) and sediment loads. They were in many respects similar to those used in Europe and America. There was a suspension bridge across the river, perhaps 250 m span, with the continuous stream of pedestrians, many carrying loads on a pole across the shoulder, very typically Chinese. There was a street market there too, one of the most interesting stalls being a medicine man's collection of weird potions and things in bottles, no doubt snake venom, mandrake and ground up rhinoceros horn! There was a temple to one of the early emperors who was renowned for his interest in water, and indeed there is one of the oldest gauges for measuring water levels in the world. Xian has now become a major tourist destination but when we were there the terracotta warriors had not long been uncovered, so we had the privilege of being one of the earliest groups of Westerners to see them. Xian itself is an interesting walled city, with several towers and minarets open to the public. We returned to Beijing by plane.

When in Beijing for the conference we stayed at the Friendship Hotel, which had been built by the Russians to house their visiting specialists. There had been almost a distinct Russian town there. It was a large hotel with perhaps 200 rooms, and dining-rooms to match. The system for taking orders and for getting a bill and paying were typical of the Chinese delight in minor bureaucracy that ensured the employment of innumerable people. Once you managed to get a waiter's eye, and then get a menu, someone else would take the order to a desk, where yet another person would take it to the kitchens – wherever they might have been – then an interminable wait until something came, then a similar rigmarole to get a bill and to pay it. The conference guests had access to large limousines parked at the front. Of course we were taken to see the tourist sites, the Great Wall being about an hour's drive away. We also went to the Ming Tombs and the Old Palace at one side of Tiananman Square.

Perhaps the most remarkable feature of the conference was our visit one evening to the Great Hall of the People for the final buffet supper. This was preceded by a meeting of the guests with a Deputy Prime Minister, for which our Swedish representative was appointed spokesperson, being the oldest delegate there. We sat in a semicircle of big armchairs. Some polite conversation was made about how impressed we were with what we had seen of the Chinese capabilities in hydraulics and sediment control, but then the female Minister for Power in effect made a public apology for allowing the Russians to make such a mess of the design of the Sanmexia project. Then the vast buffet was declared open and one must remember the privations that the Chinese had been through during the reign of Chairman Mao to understand the rush of the local delegates in their dash for plates which they filled to overflowing. They had probably never seen so much good food assembled in one place – and it was free! The conference being over, the return to Hong Kong was in the first of the jumbo jets that Chinese Airlines had recently acquired. The landing was a rough one, the pilot had apparently not yet got used to flying an aeroplane while perched in a cockpit at the height of a bedroom window!

The next trip to China was on a project that was supported by the Overseas Development Agency, to advise on a sediment problem at the Guanding Reservoir, which provided a water supply to Beijing. This was formed by a fairly high concrete dam in a valley with two arms, and it was through one that nearly all the flow from its sizeable catchment area came, with the other, much larger branch, providing most of the storage. Being fed by a river with a lot of sediment in it, the incoming flow from the arm close to the dam had silted up the reservoir locally to such an extent that the large storage volume, virtually empty of sediment, was likely to be cut off, so that the water stored there could not then be accessed. What should be done to keep up a supply to Beijing? In fact China had many different agencies concerned with rivers and water supply: Beijing city authorities; the provincial departments; the state departments; the hydraulics laboratories; the university departments; and their own hydraulics experts. There were schemes and variations on them totalling eleven that we were told about and were expected to pass judgement on because they had not been able to agree on which one to choose.

Binnies fielded a small team of engineers and hydrologists whose remit was to collect together all the available data about flows and sediment loads, which was all available to me when I got there. They had done a good job but it was soon clear that none of the schemes on the table would do more than delay the progressive sedimentation, to give perhaps two or three years' extra useful life to the reservoir. We were taken to see the reservoir in the depth of a Beijing winter when the temperature was perhaps −5 Celsius. The drive was a couple of hours, past the Great Wall where I had earlier been a tourist, out into the countryside. What were those large steel gates in the hillside that our Chinese friends didn't want us to notice? Silos for intercontinental missiles perhaps, or hardened anti-aircraft batteries? Arriving at the reservoir shore my jacket was deemed unsuitable for a walk across the ice to the other side, to see the river valley containing the main river, so I was provided with a voluminous overcoat, reaching down to my ankles, and a fur-lined cap. It certainly was cold and you may think that walking across a frozen lake is fairly straightforward because at least it would be a level surface. Not so! This was a water supply reservoir so that as water was extracted below the six foot thickness of ice, the central reservoir dropped whilst the margins became perched on the higher banks. I was lucky to have one of the locals either side of me to provide an arm because otherwise it would have been very difficult indeed. There were fishermen at work, using holes about a metre across between which they had nets under the ice, which they could manoeuvre with ropes through other holes. They loaded their catch into large sleds and hauled them to the bank.

Returning to the office provided for us in Beijing, we had first to decide what to do and then to present our findings. I had noticed that there was a deep gulley shown on the maps that crossed from the river towards the large dead storage area, and hit upon the idea of diverting ordinary flows via this gulley – which would obviously need deepening – into the upper reservoir. Flood flows, which carried the bulk of the sediment input, would continue along the original route, which could be provided with a wide, low, earth dam acting also as a wide spillway so that it would form a settling basin outside the useful reservoir area. It would even be possible progressively to modify this, raise it or repeat it further upstream, effectively providing extra agricultural land in the course of time. This was only at a feasibility stage; it would be up to the local engineers to develop it further and prepare detailed designs. We then had to present our findings to assembled Chinese experts, who sat around three sides of a hall as we expounded from the other side, being supplied with a continuous supply of tea to keep us going. After that presentation we were taken into an area with a high wall that we had seen before but had not seen what it contained. It was a hydraulic model of the dam and part of the reservoir obviously built to examine whichever of the eleven schemes we had chosen in more detail. If we had known the model was there we would not have changed our approach or conclusions, and in any case it was now of little use as it did not cover the scheme for diversion that we had recommended. It is, however, typical of a Chinese trait to keep their cards very close to their chests. A couple of years later I met one of the senior Chinese engineers who had been at the presentation and he said that our scheme had in fact been carried out. It was just a case of lateral thinking that none of the local people had seemed able to do.

My last trip to China was to Shanghai, already rather more developed than most of the country. There was more colourful clothing available than the traditional blue pyjama-like garments still worn in the less developed areas. The task was to prepare preliminary designs for a completely new sewerage system for the city, both to get rid of the serious pollution in the Whang Pu Chiang River, an arm of the Yangtze where it approaches the sea, but also to allow for future development. It was somewhat wider than the Thames estuary, quite sterile through pollution, and rather like the descriptions of the London river in early Victorian times. My role in the team was to advise on the gradients needed in the big new sewers in order to make them self-cleansing, and on ancillary hydraulic matters. The existing system was not just overloaded, it was very outdated. It had once used the night soil collection system, but when that was no longer feasible, most properties just converted to cesspits with an overflow into a stream. These were replaced by drains, but the cesspits then did not get emptied so all the foul sewage went straight into the drains, that by then got pumped into larger sewers. The pumping stations were not properly maintained, however, so that the pumps got blocked with rags and the like, and overflowed directly in the river. The necessary minimum amount of inspection of these nauseous places was carried out! The proposed system was on quite a different scale, leading to new treatment plants on the outskirts but I was not concerned with much of the project. Since then, there has been vast and rapid development of Shanghai which the scheme must have enabled. The waterfront area when I was there had changed little from the time before Chairman Mao and the Cultural Revolution, when it had been a free trade zone with Western-style buildings and warehouses.

My various visits to China had taught me a lot about the country and its culture although I saw only a relatively small portion of this huge and very varied land. The Chinese have always had a skilled engineering profession and can hold their own in any technology. It is difficult to imagine our multi-party system of democracy working there, yet clearly their one party toe-the-line system had led to terrible hardships in the past.

CHAPTER 12

South East Asia

THIS AREA, SO FAR AS MY INVOLVEMENT is concerned, goes in a great arc from Thailand, through Malaya and Singapore, into the Indonesian islands of Sumatra, Java and Sulawesi, with Papua New Guinea further to the east. Indonesia itself stretches over almost 4,000 km, with several time zones, and many active volcanoes. I will start my journey in Thailand, one of the earliest places I went to on business. This was to a hydraulics conference in Bangkok, where I stayed at the Dusit Thani Hotel. Strange that the name of this hotel has stuck in my mind from the dozens I have used over the years. It was then the highest building in the city with about eight storeys, but by now either demolished or dwarfed by others. Bangkok is today very much on the tourist map, and there are interesting places to visit, such as the royal palace and the floating market. The conference delegates did a city tour on a day off, but the things I remember best were the folk performance in the hotel as I dined one evening, a couple in ornate gold costumes dancing elegantly to the music of a gamelan, a percussion instrument involving bells, gongs and cymbals, then being by the Chao Phraya River when a squealing pig wrapped in a basket-weave strait jacket was brought ashore, no doubt to go to the butchers, and then a trip up into the hills to a rather primitive village with working elephants – and lots of locally made knick-knacks for sale! En route, we called at the River Kwai, site of the famous – or really infamous – bridge that British prisoners of war were forced to build as part of the Burma railway line where so many of them died from sickness and Japanese brutality. It was one of the most moving experiences of my life to walk over the railway bridge (not actually the wartime one) and think that for each sleeper I stepped on one British prisoner of war had died. The Japanese delegates to the conference were conspicuous by their absence from this trip.

Later when Sheila lived in Hong Kong Margaret and I took a holiday with her to Bangkok and the Thai resort of Phuket, before

the time when it became somewhat overdeveloped. Wonderful
tropical seas to bathe in, a hotel right beside the sea, a boat trip to
a nearby island where the locals would barbecue a large fish for you
for lunch – but a painful walk over coral to get back! We also did
a trip to a timber-fishing village built in the sea on stilts, and an
island where an early James Bond film had been made. The
Bangkok conference was, however, my only business trip to the
country.

Proceeding south down the peninsular is Malaya, the location of
many professional visits over the years. Sometimes it would be just
a drop off to or from Hong Kong, for minor consulting work for
the office that Binnies then had in Kuala Lumpur. It was in a block
of offices about six storeys high, and one of the most noticeable in
the then town centre as it had a diagonal lattice of steel on the
outside as part of its structural design. We were much involved
with the development of many regions of the country through the
provision of water supplies. There were several on the east coast
where the problem was salinity intrusion. This occurs where a river
joins the sea, with the rise and fall of the tide having a piston-like
effect on where the fresh water gives way to primarily salt water
from the sea. Any intake for drinking water has to be upstream of
this limit of salinity intrusion, which depends on whether it is
spring or neap tides and the flow in the river. Any period of
drought is critical in that one needs the water, yet the salinity
intrusion is at its worst. We used a mathematical model developed
at HR Wallingford, which allows for the mixing between the two
bodies of water as well as this piston effect to decide just how far
upstream any water intake would have to be. Water supply also
involves reservoirs and hence dams.

On one of my visits I joined the local chief engineer on a trip
into the forest to inspect the site for a new dam. The choice of
location was a matter for geologists rather than a hydraulics man so
my only input was on the best form of spillway, but the trip
included one extraordinary encounter. Deep in the forest on a dirt
track accessible to our Land Rover and similar vehicles we heard
the tinkle of a tune just like an ice-cream vehicle's signal. Lo and
behold, a Walls 'stop-me-and-buy-one' motorcycle and sidecar
came into sight, a vehicle just like those familiar to me in my

childhood. What was he doing in the depth of a Malayan jungle? Chatting to him, it seemed he made this journey once a week to find where the loggers were working and to do some business with them. We enjoyed our unexpected ice-cream anyway.

A few years later, son John was resident engineer on this same dam, and so when we holidayed with his family in Kuala Lumpur I was able to visit the site again, now cleared of its trees, with the earth dam well on the way to reaching its full height. John had a rented house not far from the site, whilst Lizzie and the boys lived in a block of flats overlooking Kuala Lumpur. This had its swimming pool which the family enjoyed and we took a swim there on one New Year's Eve before catching the flight home with wet togs in the case. On that trip we also had one of those international contacts that are so surprising at the time. Coming up the escalator in a department store, we came face to face with a Japanese professor, Prof. Iwasa, who was giving a course of lectures at the local university, and who had also been on the Council of IAHR. The boys went to a school for Europeans there, and John got home to them each weekend. Even then, Kuala Lumpur had changed almost out of recognition but since it has become a symbol of Malaysia's development with the Petronus Towers, for example. Whilst there we took a train journey down to Singapore for some Christmas shopping and sightseeing.

Binnies also had an office in Singapore where the partner in residence, a Mr Hooi Ka-hung, was a local engineer of Chinese extraction, so once or twice I had a short stopover there in case there were any local hydraulic problems requiring advice.

One of my trips to Malaya was effectively a short stopover fitted into a return trip from Hong Kong, when I was asked to call in to advise on a beach erosion problem in Penang, at the Rasa Sayang Hotel. Their grounds came right to the beach and recent erosion meant they were worried about how to protect their frontage. My local contact could speak Malay of course and I got him to enquire of local fishermen whether there had been any unusual pattern of weather. They confirmed that recently there had been a predomi-nance of strong winds from the south-east, whereas they normally blew from a more northerly quarter. The bay at Batu Ferringhi, which translates as 'the foreigner's beach', is a typical horseshoe several miles in length, enclosed by headlands at each side. These trap the sand but if strong wave action occurs not directly on to the beach, the sand will shift from one side of the arc towards the other, and this is just what had happened, the Rasa Sayang being right at the northern edge of the arc. I could advise them to do nothing, to let nature take its course, when a more normal wind direction would bring the sand back to their frontage. Above all, they must not build a steep sea wall, because this would surely cause beach erosion in front of it. In due course the tourist brochures for the hotel showed that its beach had indeed returned.

The Straits of Malacca separate Malaysia and Singapore from the westernmost island of Indonesia, Sumatra. It is indeed an island but a big one, over 1,500 km long and 400 km wide (1,000 miles by 250 miles). There is a mountain range to the east side but the west has a much flatter landscape where there are rubber plantations and palm oil farms. Once on a holiday trip to Hong Kong to visit the family there was a call to 'hop over' to Sumatra to advise on an irrigation project there, a flight of about 1,200 km. As Margaret was with me I agreed only if she could come with me and this was about the only business trip, apart from conferences, that Margaret was able to come with me. We flew into Medan but the job was located some 250 km to the north near Banda Aceh, the journey there by Land Rover being one of the most uncomfortable I have ever done, over potholed dirt roads. We stayed with the local representative who had a bungalow near the sea. He had a fine collection of museum-quality shells – and a lot of mosquitoes to keep us awake at night, which the toad in the corner of the bathroom did not seem interested in! Margaret had been intrigued when visiting the beach one evening to see the young people bathing, the girls in this Muslim country keeping their dresses on when in the water.

The job itself was quite interesting, involving the refurbishment and enlargement of an irrigation scheme, which had become blocked with sediment from the gravel bed river (Fig. 25). The solution was to move the intake to a better location on the outside of a bend, to feed a tunnel through a rock ridge that formed the river bank, into settling basins on the other side where there was space for them. This was the only site inspection of my career done with an armed guard, a local man with an old rifle. There had been some murders of American oil workers elsewhere on the island – an early example of Islamic terrorism perhaps. That weekend, it was arranged that we should spend our spare time at the island in Lake Toba, an ancient supervolcano with a lake in the caldera 100 km long by about 30 km wide, an island of about 20 km long in the middle. It was a long drive from Medan to the lakeside, where there was a boat waiting to take us across the lake to the one hotel on the island. It was a bit of a tropical paradise, remote from anywhere else although the island was large enough to have a village with a market that we were able to walk to one day. Food at the hotel was a bit unimaginative, chicken for dinner with pineapple for dessert, and pineapple again for breakfast. We got rather fed up with pineapple! Monday morning and we took the boat to the shore to await our transport, and as it was a long wait sat on our suitcase, but the driver did not let us down. So it was back to Medan and ultimately home. Banda Aceh was destroyed by the tsunami on Boxing Day 2004. Did the bungalow we stayed in survive? I doubt it because it was only about 30 feet above sea level as I remember.

Going east from Sumatra is the island of Java, with the capital city Jakarta. This is part of the Pacific rim of volcanic areas. It is intensely populated, with terraced fields of rice. There is a high rainfall but nevertheless many areas are irrigated, and my professional activities there have included advice on river stability and irrigation intakes. This meant a train journey of several hours from Jakarta, getting off the train at Purwokerta to meet my contact and go to the River Serayu. There is limited opportunity for internal flying in what is basically a mountainous country and train journeys involve routes winding their way through passes and tunnels to get over the mountain spine – and down the other side. One such journey was between Jakarta and Bandung, a city at a high elevation thus with a very pleasant climate despite its proximity to the equator. This was when flights on that route were regarded as a bit dicey. The tale was that the state airline, Garuda, used fighter pilots who sometimes forgot they had passengers aboard – but it was more likely that risks were taken when landing in bad weather. Bandung is a university town, and the location of the major hydraulics research station there, staffed by engineers who really knew what they were doing.

There was an Asian conference there that I attended, and during the first night we were there Mt. Merapi had one of its periodic eruptions. Waking up early, I saw a very red sun creep above the horizon and then disappear as it rose further, creating a period of dusk. Looking out of the window, everywhere was grey, the tops of the lampposts even had a cap of grey 'snow' about 2 cm deep. This was the fallout from the huge quantity of fine material that had erupted, a surprising material in that it was of non-cohesive single-sized grains, about a tenth of a millimetre in diameter (Figs. 26 & 27). Conference-goers were issued with straw hats to protect their heads against any further fallout – and I still wear mine when the weather here calls for it! This eruption was more or less a repeat of one that had almost caused a BA jumbo to crash. On its way from Singapore to Australia it had run into a similar high level dust cloud and all four engines stopped. Jumbo jets are not built for gliding or for landing without any power but the efforts of the crew managed to get one engine started as it plunged down. They had put out a mayday call of course and the airport at Jakarta had cleared the decks for an emergency landing. Just before getting to ground level, the crew got a second engine started at reduced power and this was just sufficient for them to come straight in and land. I don't think the crew were ever rewarded for the skill and bravery but their actions saved many lives.

The people in Java have had to live in close proximity to volcanoes and one of the problems they face is the risk of lahars. These are flows of a mixture of sediment and water that really are more like giant mud-slides. The sediment may include large rocks and in these phenomena they float down the valleys from the volcano summit on this dense mix of finer material with the water which may come as rain but which may itself have been ejected from the crater. These flows can cover, and hence destroy, any agricultural land although in the course of time they mature into highly fertile ground again. There are many works in central Java to try to control these flows, for example by building dams to prevent them getting access to side valleys or by building a long bund in the form of a loop where they might emerge from the end of a valley to spread more widely over the rice fields. These are called sabot dams, from the Dutch word for a boot – they have the shape in plan of a footprint. I was able to inspect many of these schemes, including successions of dams built across valleys to catch the heavy movement of sediment that later rainfall will generate down any of the volcanic valleys. It was a case of me learning about the problems that others face rather than giving any advice!

Sulawesi is a strangely shaped island to the north-east of Java, still part of Indonesia. Binnies were brought in by the Overseas Development Agency to advise on a big new irrigation project there. In essence, Java was overpopulated and its increasing population needed space to expand into, and in which to grow its own food. There was a plan therefore for opening up Sulawesi to the excess population of Java. This involved a flight of about 2,000 km (1,200 miles) from Jakarta to the only place there with an airfield, Ujung Pandang, which used to have the name Macassar – where Macassar oil came from and the origin of the Victorian need for antimacassars on the furniture, as Macassar oil was used on gentlemen's hair! Pare Pare some 200 km away was my final destination, but it was not easy getting there! The office car was due to pick me up from the Jakarta hotel quite early but I was awakened by the sound of very heavy rain. That was before the new international airport was built but it was a domestic flight anyway. The driver was there in the lobby waiting for me and was in a hurry to get away because he guessed there would be floods and indeed there were. That area is extremely flat, which makes it difficult to drain but meant that the flooding was widespread, varying from perhaps 15 to 45 cm deep, enough to stop a car's engine if it got into the exhaust pipe. The car was a Mercedes so pretty robust but many vehicles broke down and had to be pushed out of the way by youths who were charging for the service. The driver explained that they would put their flip-flop sandal over the exhaust if you weren't careful, so stopping the engine and claiming their fee! We got to the airport at about the time the flight was due to depart – but the airport was flooded anyway. At the departure building it was a case of paddling through 15 cm of water to get into the building where there was still water on the floor. Checking in went ahead despite the problems and within the hour the runway had drained enough for the flight to take off.

Getting to Sulawesi involved several stops on the way, at Semarang and Surabayu then across the Java Sea to Banjarmasin in Borneo (also part of Indonesia) then across the Makasar Sea to Ujung Pandang. However, as we got near our destination the heavy thunderstorm had got there first and we flew around for the best part of an hour waiting for it to pass. The airfield had no modern aids so landing was only possible in good weather. The pilot must have realised he was getting low on fuel so we returned to Banjarmasin, where he was able to refuel. There were no hotels there for the passengers still on board so we returned to Jakarta and I was back at the hotel, still with wet feet, at about 10 p.m. having left at about 7 a.m. I had checked at the airport that I would automatically get on the next day's flight, so it was a repeat procedure. Somehow the driver had got the message that we needed to try again and was there to collect me. The flight went smoothly, and my contact working for ODA was waiting there for me. The previous day he had made the trip from Pare Pare, braving the floods, only to have to return and try again the next day. My contact there was none other than Leonard Watkins, whom I had worked closely with when he was at the Road Research Laboratory and I was at HR Wallingford, on the programme of research into rainfall and run-off. He was now an ODA employee. Another experience of the small world!

Pare Pare is very close to the equator and Sulawesi is a tropical paradise. The town was mainly one street with bungalows on either side, but it was rather basic. There was a petrol station, selling the fuel from drums lifted to pour the fuel into the tank via a funnel. There had once been a hotel with a swimming pool but that was derelict. The only 'restaurant' specialised in a sort of pancake, cooked on a hot plate of steel with a wood fire underneath, and some sort of filling inside. I was staying at Leonard Watkins' place and he had a local girl as servant, so it was fairly comfortable. The project was a study of a complete irrigation scheme, with a dam where the river left the more mountainous area to form a reservoir, a barrage across the river to provide the head pond from which water could be taken on both sides to feed main irrigation canals and settling facilities to stop the coarser sediments from being carried into the irrigation system. It was my job, following a full inspection of the site, to advise on the layout of the irrigation headworks and on the design of the individual components. Of course Binnies would supply the engineering staff to fill in all the details, carry out any necessary surveying and do the design drawings. Several years later, when the scheme was complete and more people had moved into the area, I was asked to visit the site again, to give further advice on the operation of the sediment removal facilities. The whole scheme seemed to be performing just as expected. This was a gravel bed river and this material moved during floods, of course. It had built up to near the crest level of the weir forming the barrage, but this was as expected (Fig. 28). There was no untoward scour downstream either. The area was now under cultivation and Pare Pare had improved somewhat, with a petrol station that had pumps!

The Indonesian people are very hospitable – once they establish that you are not Japanese! They have bitter memories of the Japanese occupation during the war. We took a weekend trip into the hills, known as the Tana Taraja region. Accommodation was in a bungalow hotel with perhaps six rooms, rather surprisingly called Rose Cottage. It was indeed a bit like an English country cottage and had roses in the garden, with a beautiful blue sky above. This was our base but we went from there to see a local village, where the long huts were made entirely of bamboo and there were rice barns, looking a bit like upside down Noah's Arks, set on stilts to keep the vermin out, with very overhanging roofs dipping down in a gentle curve to the middle (Fig. 29). It really was a fascinating place, and the people there bury their dead in clefts cut into the walls of quarries or other cliffs, with colourful and clothed effigies in balconies above them (Fig. 30). Of course European visitors were quite a novelty then but now these places have got onto the tourist trail and no doubt Rose Cottage has been demolished to make way for a multi-storey hotel. The last morning we were there must have been market day because, after breakfast, a bus appeared to pick up more passengers from the clearing in front of our little hotel. The buses in Indonesia are small affairs, perhaps basically twelve-seaters but they can squeeze many more of these rather small people on them. It presented an amazing sight! Bamboo is used for many things and I discovered an unexpected one, for carrying the local beer. These bamboos were about 20 cm in diameter, hollowed out and stoppered at the bottom but open at the top, and about 2.5 m long. Clearly they would not fit into the bus so they were carried with their tops out of the windows, so that the bus looked like a mobile organ, with its range of pipes foaming! A most extraordinary sight but unfortunately I did not get a photograph.

The next large island to the east is half part of Indonesia, West Irian, and half largely administered by Australia: Papua New Guinea. PNG has a spine of mountains reaching heights over 2,000 m and is covered with dense tropical jungle. It has an extremely high rainfall, many metres in the year. The capital city is Port Moresby, with the only international airport. Appropriately enough because of its Australian connection, the project there was being carried out by our Melbourne office, so my visit was from there, rather than from Indonesia. We were to assess the impact of a proposed gold and copper mine on the adjacent river valley. It is a strange fact of geology that significant deposits of these minerals occur at the tops of mountains and of course it is the geology profession that has the primary input into such projects. However, any mining operation produces a great quantity of waste rock, some from excavation to uncover the mineral-bearing seams, some from the crushing and initial processing that gets carried out on site. There were concerns about what would happen to this material when it got washed into the river system, including the effect of any raised bed levels on the few fields, referred to as gardens, that the sparse local population depended on.

The site was only accessible by air as the jungle was impenetrable, so after a night in Port Moresby we caught the private plane operated by the mining company. This was a twin-engined plane about a ten-seater and it was a flight of several hours' duration, mostly over the sea, skirting any tropical thunderstorms which would generate high turbulence. The landing strip at the far end was in the foothills of the mountain and we turned north from the Coral Sea at the mouth of the Fly River, the flood plain of which is just a vast mangrove swamp. We had dropped below the clouds and the pilot turned left up the valley of the Ok Tedi, the river we were concerned with. He had radioed ahead to check all was clear for a landing and as the valley sides closed in we realised there was no escape – he would have to land or crash! Fair enough, suddenly a grass clearing on the flank of the valley appeared and we were down. The airport building was a corrugated iron-roofed garden shed, but there were a couple of vehicles there waiting for us. The camp was a wooden building with a corridor at one side linking the couple of dozen rooms, each with two berths, and at one end the canteen area and some showers at the back. This would be our home for about a week, myself and Larry O'Dell, an Australian engineer from the Melbourne office, who would be responsible for any local input when carrying out the studies after our field trip.

Although there were some very local tracks joining up various work facilities, access to the mine location up in the mountains above us was only by helicopter, as indeed was access to the river below us. Helicopters could only fly when the weather was right in this mountainous terrain, but the weather pattern at that time of the year was reasonably well-established, with rain mostly at night. Our trip up to the site of the proposed mine meant flying just above the tops of the trees, in the gap under the thick clouds, to a shelf formed from rock that had been excavated in digging a tunnel into the mountain side so that the geologists could take samples and see the rocks much better than they were able to from the many boreholes that had also been drilled. The platform was too small for the pilot to land safely and he was not going to hang around, so he hovered about two metres above the ground while we clambered in turn on to one of its skids and jumped down. It was agreed that he would return at 5 p.m. We walked into the tunnel with our helmets and torches to help us see the rock around us and were back at the ledge outside in good time in case our transport came early. I then realised just how much colder it was at that height even though we were just five degrees south of the equator. Should I not have brought a jumper in case the weather turned against us and we were benighted there? Luckily all was OK – exactly at 5 p.m. a small dot appeared below us and we heard the sound of the helicopter as it came up the mountainside skimming the trees. Again he did not land but we scrambled aboard as it hovered and down we went to the landing strip and back to base to have a shower and then go along to the canteen for a beer before dinner. This was an American operation, so food was pretty good even though it all had to be flown in. The canteen was very pleasant, with bougainvillea around the railings and in the awning above, and later in the evening there was the twinkle of glow-worms.

One day, after we had finished our task for the day, two walkers appeared. They had trekked for two weeks from Port Moresby through the jungle. The young woman just said, 'Where's the shower? I need a good wash.' Her legs and arms were purple-blotched from the gentian violet she had used on the mosquito and leech bites. Her male companion was a local guide, a typical Papuan, naked but for his penis guard made from a gourd and tied in place, with a bone through his nose, feathered bangles on his arms and his string bag over his shoulder containing all his possessions. Did you know that the string shopping bag was invented by the Papuan natives? Actually the natives had until recently been head-hunters and have taken slowly to civilisation, learning to speak pidgin English, putting their spears away, and now more normally wearing a T-shirt and shorts, but it was intriguing to see one in native dress – or rather undress!

Our next field trip was to the river itself, primarily to look at and take samples of the sediment in the bed. This meant awaiting the availability of a helicopter which could land on the shoals in the river (Figs. 31 & 32). We worked downstream and had polythene bags for our samples. The usual way to sample bed material in this sort of river is to walk across a selected section, and pick up the pebble under your right big toe. When you have about fifty, then you have a valid sample of the grading of the bed material, which is what one needs to know if you are going to assess sediment movement. We also needed to know how much and in what sizes material would get released from the deposits from the mining operation, and here the geologists and mining engineers had all the information we required. There were many kilometres of borehole samples kept in row upon row of trays in the borehole storage hut, which we were able to see (Fig. 33). One of the standard measurements the geologists made was the fracture index, in effect what size of piece the rock prefers to break into

35. Sidney: the Opera House which was an engineering challenge; the Harbour Bridge behind

36. En route to the tunnel through the Andes, Peru

37. Exit portal of the Transandean Tunnel

38. Son David's project in Lima: curing arch sections for suspended floor; roof sheets to left

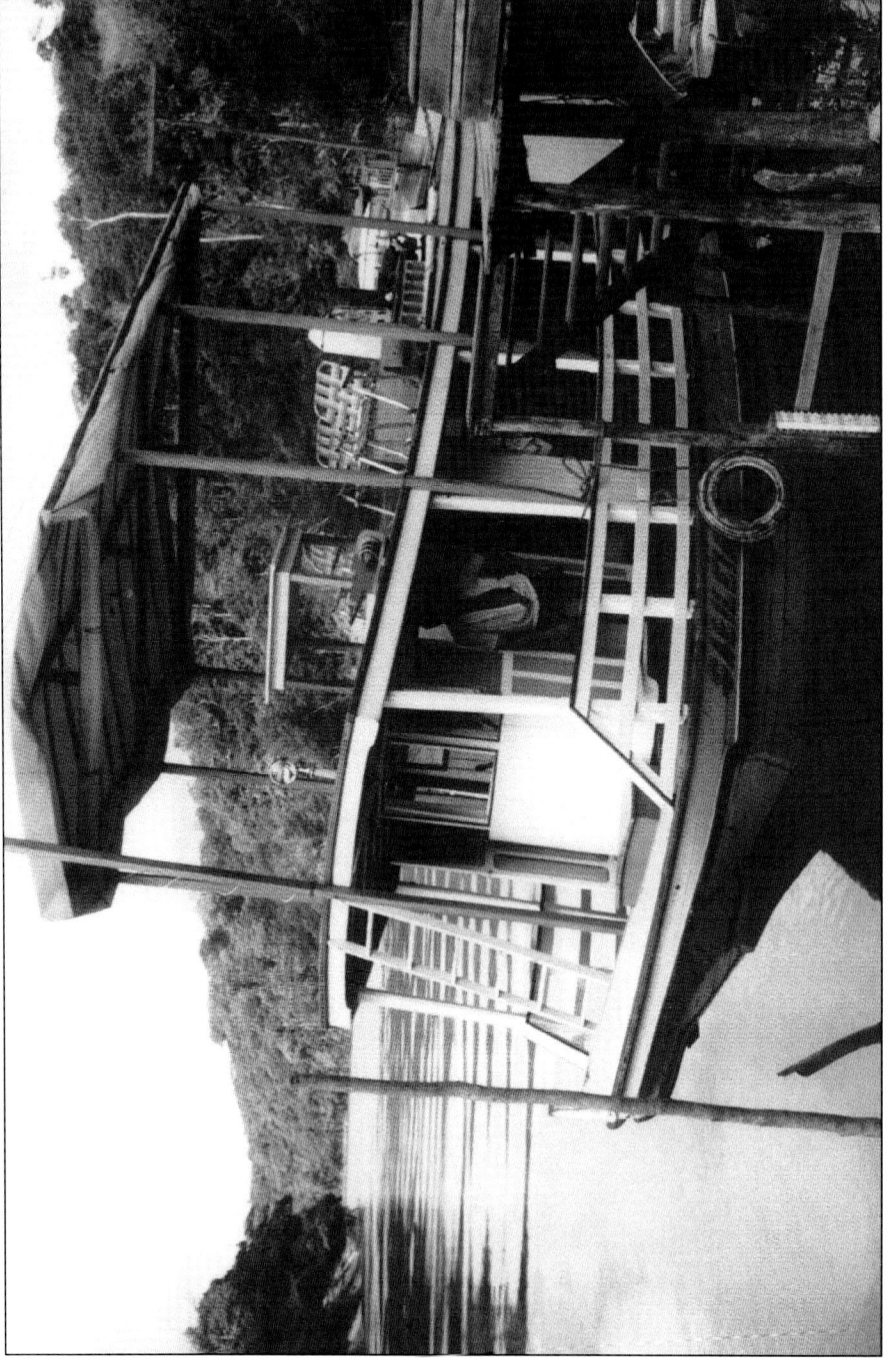

39. Our boat for the river inspection in Brazil

40. Water level gauges on the River Capim, Brazil

41. Sediment removal lakes on the River Tame

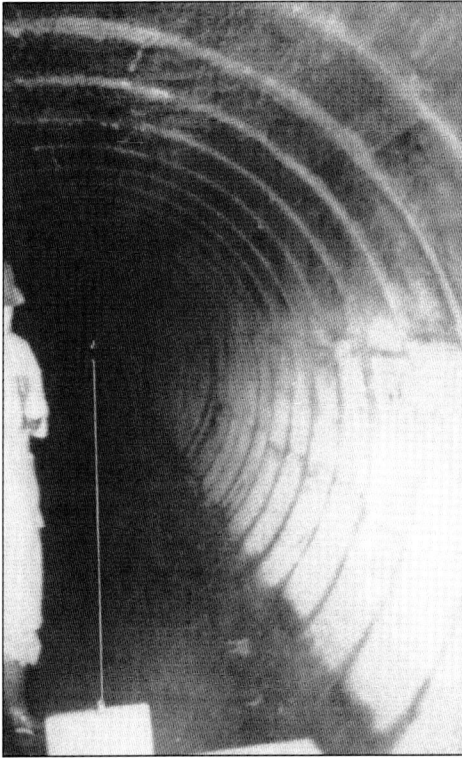

42. *Laser equipment in use in research on resistance of segment lined tunnels*

43. *The Severn Barrage proposed location, with islands Flat Holm and Steep Holm*

44. *Sectional detail of proposed turbine caisson*

through any natural cracking. They had also identified various rock types, and these varied in hardness depending on where they were in the mountain mass. So the mining engineers, predicting how much of every sort of rock they would dig out month by month during the life of the mine, could tell us what amount we had to allow for being washed out of the various dumps into the river valley.

Our work on site completed, we took the first available flight back to Port Moresby and so on to Melbourne. My colleague, Larry, had an aversion to seat belts so would not actually fasten the one in the small plane in the first leg of our return journey – he just pulled it across to make the pilot think it was fastened. But small planes are very subject to turbulence which is especially bad in the tropics. The plane could hit an upward or downward current at any time despite the pilot's efforts to steer round any bad spots he could see ahead. We had several such jolts and Larry finished up banging his head on the roof of the little cabin, though fortunately this was only about three feet above us! Arriving back in Melbourne, the immigration officers were not going to let us take our sediment samples into the country without them being sterilised, so we had to give them up at the airport, to be collected later. Whilst this discussion was going on the baggage had arrived at the carousel. Larry's was a large tartan suitcase easily identified and as I looked ahead I saw a young woman collect it and head for the exit. What was she up to? I told Larry and he rushed after her saying 'You've got my bag!' When she checked she found it was indeed so: hers, surprisingly, was identical.

Back in Melbourne we could get to grips with the task in hand, of predicting what the effect of the mine would be on the River Ok Tedi. The first job was to work out how the different rock types would abrade as they travelled downstream. It is a matter of common observation that rivers are very steep in the mountains with coarse material in the bed, going from boulder sizes down to cobbles and gravel. As the river goes through the transition to the plains downstream, the coarse gravels progressively give way to finer gravel, then more coarse sand in the mix, down to fine sand and finally to silt. This is due to the interaction of the progressive abrasion of the pieces as they rattle their way downstream and the

inability of the slacker gradients to convey the coarser fractions of
the grading that the steeper upstream reaches could. The familiar
concave upwards curve of the river profile results. We knew that
some of the rock we were concerned with was very friable, so
would easily break down, whilst some was much harder. The way
we looked at the abrasion was to take samples of perhaps four
different types of rock with a grading down from cobble sizes, and
subject them to a standard form of testing used for road stone. The
technical college in Melbourne had the appropriate equipment
which consisted of horizontal drums that were rotated by an
electric motor so that the material in them tumbled over as it fell
down the rising side of the drum. In our case the test was done
with water in the drum as well and the results, when they emerged
in the form of before and after grading curves, could be expressed
quite simply as if everything was abraded from the coarsest grade
and finished up in the finest fraction. In fact, what happened was
that the bits knocked off the larger lumps moved down in the
grading curve but the parent-size itself would in due course move
to the next size below, a sort of cascade effect. The results,
however, were simply expressed as a loss in weight per tonne per
km of the drum's peripheral rotation, moved from the coarse end
of the grading to the fine end, an easy concept to incorporate in
the computer model we were to make. Actually, the preparation
of the program was done in the London office where we had the
same sort of Hewlett Packard desktop computers that were
available in the Melbourne office. This was long before the advent
of the modern personal computer, when the screen was just a green
listing of the program, which had to be fed in via cassette tapes.

I had very good computer chaps in my team, especially Graham
Thomson who linked together all the different components: the
flow regime over time as provided by the hydrologists; the rock
types, gradings and dumping programme supplied by the mining
engineers; the geometry of the existing river channel from the
survey we had requested; the existing bed material grading from
our site work; the abrasion characteristics of rock from the
Melbourne tests; and the Ackers and White sediment transport
function adapted for graded sediments. It sounds very complex but
the basic equations were quite simple so it was just a case of

number crunching, and we were only concerned with the rise in bed levels over the various sections of the river over time. The model was run in Melbourne and although a modern computer would make short work of the computations, in those days the program and data would be put into the system one day and run overnight, taking perhaps five hours to complete its task and print out its results on a line printer, a normal form of output in its day. So with this work we were able to make our forecast of how the river would develop. The results seemed very convincing and the native gardens were not at risk but we were never able to check them properly when the mine came into production because there was a change in the market value of gold and copper which meant that the scale of operation was much reduced. This had been one of the most interesting jobs in my long career involving a deal of new methodology – and I had seen yet another remote part of the world too.

CHAPTER 13

All the capitals down under

M Y FIRST TRIP TO AUSTRALIA was on the way to New Zealand to inspect the Waikato River while I was at HR Wallingford, and it was in the days when the Boeing 707 was the main intercontinental airliner, with much shorter range than today's jumbos. It therefore required several stops and my recollection is that we called to refuel or pick up passengers in Cairo, Bahrain, Bombay, and Singapore before heading south over the great sandy wastelands of Australia, en route for Sidney. It was an overnight stop in Singapore and we saw the dawn break over this vast continent. That impression of the size and emptiness of Australia has stayed with me. Other trips there have been to assist with work being undertaken in the Melbourne office of Binnies, or involving the firm's representative in Perth. However, an early trip was with Stanley Ford, the partner who looked after interests in Hong Kong, and it was a very unusual project he had got involved with. The ocean liner *Queen Mary* was being converted (after its retirement from the Atlantic run) in Hong Kong harbour to become a floating university, when it caught fire. The water pumped into it finally overcame its stability and it rolled over and sank in the middle of the harbour. It could not completely overturn as its superstructure hit the bed, and so it was mostly on its side, with the uppermost side sloping at about twenty degrees, and its superstructure mostly under water. It was thought that as Binnies had an office there, we might have a good chance of getting the contract to right it and refloat it: there was good money in salvage.

It was a very strange task for civil engineers to get involved with but when you think further, we understood the hydraulics of flotation and stability, had people with experience to design anchorages and hawsers for pulling it back on its bottom, the firm had experience of dredging which would probably be needed to release it from the mud and the rest was a case of holding it steady whilst it was pumped out. There was little damage to the hull. So why the visit to Melbourne? It was to discuss it all with an international salvage expert who lived there and whom we hoped to associate with when making a bid. We had first visited the 'old lady', walking on her flank, now a little rusty but solid enough, the only time I have stepped foot onto an ocean liner! The thing I remember about visiting the salvage man in Melbourne was that he had many trophies from his salvage work, such as a golden model ship given to him as a bonus for salvaging a Japanese tanker and, getting it from the safe, a Tompian clock, the most valuable timepiece of the antiques world, and even then worth some £60,000. However, the *Queen Mary* was never salvaged: it was cut up underwater and the pieces taken for scrap.

Other Australian visits were for more conventional civil engineering. One was the design of huge new sewage outfalls for Sydney in the Botany Bay area. This would include a tunnel under the sea with vertical pipes from it called risers, culminating in a group of outlet ports. Provided these were in deep enough water, the buoyant plumes from them would undergo turbulent mixing with the surrounding salt water, become diluted and swept away by the tidal currents. So there were a number of issues to be settled. What depth of water was needed to give, say, 100 fold dilution? How far apart would the outlets need to be to avoid overlap of the plumes? How strong and in what direction were the tidal currents? All these had to be factored into the many other design parameters. Another visit was to do with the problem in PNG already described, and then Larry O'Dell and I visited Alice Springs to meet an environmental expert who was also to advise on the effect of our sediment predictions. He lived well into the outback breeding horses, but the feature of Alice Springs, a modest town of mostly one- and two-storey buildings, was that it rained, and the smell of the eucalyptus trees was quite powerful. Rain there is unusual, but we did not experience a flood like Prince Charles and Diana did on their honeymoon!

Binnies were leading consultants in the study of the Severn Barrage project, but it meant that the firm had the experience to advise on tidal power anywhere in the world. One area with tides almost as high as the Severn estuary is the north-west coast of Australia, the area of Collier Bay. This meant a Qantas flight to

Derby, where we picked up a private hire plane to take us further north to overfly the area, especially Secure Bay, which was the most likely choice of location. It was the time of spring tides and the timing of our flight was such that we saw the main ebb of the tide after having filled the region behind the narrow entrance. This was flanked by rock cliffs. Only solid rock could withstand the tidal currents of several metres per second. It was most impressive, but what a remote place! (Fig. 34) There are no roads, it is rocky desert, just fifteen degrees south of the equator, with mangrove swamps at the margins of the tidal areas. We returned to Derby before the fuel ran out and found a small cafe for some lunch. On the way to and from the airfield, the problems with the indigenous aborigine population were all too clear. The roadside verges were littered with discarded beer cans, evidence of how much of their social security payments was spent on booze as they went back to their homes. A sad sight, really. For us it was back to our plane and north again to see the next flood tide come in: an equally impressive sight, with strong vortices shed from the lee of the cliffs on either side.

Despite the severe conditions, we were confident that a tidal power scheme could have been built near the mouth of Secure Bay. The cliffs would provide anchorages for cables to manoeuvre caissons into place, some containing gates and others the turbines. In the event, soon after our feasibility report was submitted, gas was discovered offshore and the energy needs of the area were amply met, with some to export, when the gas field was developed. In much more recent years, a tidal power scheme has indeed been built there, but of more modest scale, in a different position and using adjacent reservoirs, alternately filled by the tide, with the generating capacity between them.

Margaret was able to join me on our visit to all the capitals in turn. This was a lecture tour organised by the Institution of Engineers, Australia, and funded by a major manufacturer of asbestos cement pipes. For me this involved giving an evening lecture to the assembled engineers, having spent the earlier part of the day visiting the local university and hosting a seminar there. This annual lecture series helps the Australian profession keep in touch with what is going on in the rest of the world, but it involves a tough task for the visiting lecturer, bearing in mind the flights from city to city, with the time changes involved. We were well looked after in terms of hotels, always the best, and the ladies both looking after Margaret during the day when I was busy, and seeing that there were bouquets of flowers to welcome us on our arrival. We went from Sydney, to Adelaide, Melbourne, Hobart, Canberra, Perth, Darwin and Brisbane so we saw not only all the capitals, but were also able to visit botanic gardens – Margaret especially became quite an expert – and enjoy much good hospitality on the way.

Arriving in Sydney after a long flight, we were taken straight to a barbecue dinner arranged at the Manley Vale laboratories, one of the major hydraulics research institutes in New South Wales, but it was raining so that it had to be held inside a large hall, more like a hangar with open doors. Very pleasant, but we felt more like going to bed. Manley Vale is across the harbour from Sydney, and our hotel room there had its own kitchen and we had a view direct to the nearby harbour bridge, a wonderful example of British engineering (Fig. 35). It is very pleasant trip by ferry across the harbour to Manley, and close to the bridge is the old part of the city, now dwarfed by high modern development. Not far away is the opera house, a fine example of architecture that was almost impossible to build even though it looked impressive on paper – and not too well suited to its purpose either from what one reads – so there were high cost overruns. Sydney is a fine city, with extensive botanic gardens. We were taken on a picnic on our weekend in Melbourne, again with a barbecue meal, being watched by emus, one of which nipped in smartly and stole a steak off the embers when we weren't watching. They look fearsome creatures in the wild with their large beaks just at eye level!

Next stop was Hobart, capital of the state of Tasmania. This has a temperate climate, being well south, and Hobart is a smaller city than the other state capitals. On arrival I was asked to meet a local reporter who had obviously been briefed that there was a visiting civil engineer from the home country who knew all about dams. There was a proposal to develop hydropower in a valley in the rather pristine landscape of this very wooded country and this was very contentious, because the local environmental lobby – and many from further afield – objected since it would despoil virgin natural landscape with rare trees that were survivors from ancient forests. Of course I declined to express an opinion as I knew nothing about the project. Hobart has many old buildings and then had no skyscrapers. On the promenade by the harbour, there was a parade of veteran cars that entertained us during some time off from my lecturing.

The next stop was Canberra, not a state capital but the capital of Australia as a whole. Its character is quite different in that, being created as a national centre of administration, it is very spacious with lots of green, but I think many Australians think it is a somewhat boring place compared with the much more commercialised cities, which also have more cultural activities. Most of the time there was occupied by the usual collection of lectures and seminars, and again Margaret was looked after by the ladies including a visit to a vineyard with much sampling of the local wines, especially by her hosts.

An excursion from Canberra was to Cooma, which was the headquarters of the Snowy Mountain Authority, responsible for the construction and operation of a whole complex of hydro-power installations, with dams, connecting tunnels and power stations. We were their guests and we joined a flight in a small private plane they owned, about a five-seater, to visit several of the installations, landing on grass strips, and seeing the scheme from the air. Margaret enjoys trips in small planes, so this was quite a thrill for her too, seeing things from a low level rather than from 30,000 ft. One of the lunches I do remember was during the Canberra visit, at a restaurant run by an ex-test cricketer, where the waitresses wore short skirts – and very little else – an experience I have not come across elsewhere, but the food was OK and obviously a popular eatery for the Aussie males!

Perth is a very fine city far removed from the rest of the developed areas of Australia. There was the usual round of lectures and seminars while Margaret visited yet another botanic garden, but a weekend break gave us the opportunity to take a trip by boat from the port, Freemantle, to an island which is the only habitation for *quokkas*. These are marsupials that spend most of their time on two legs, about the size of a cat and very tame in that they expect visitors to feed them with food bought from one of the stalls on the island. These animals are a perfect example of what happens when a species develops separated from other species and with no predators. Also while there we were invited to a barbecue close to the inner harbour, so it was possible for some of the party to bathe – but we were warned that there were nasty fish lurking just under the surface of the sand that poke their top fin upwards – and this was poisonous, giving very severe pain. We kept out of the water! The flight from Perth to Darwin was overnight, and very tiring being in a fairly small plane with several stops en route to keep us awake. Darwin had suffered its severe hurricane a year or two before – and it was stewingly hot! Our host took us into the nearby jungle to a pool where it was safe to bathe, to cool us off and it was indeed very pleasant. Without this local knowledge bathing in any inland river or lake would be very unwise, not only because of the alligators but also various nasty snakes, and water bugs. Next stop was Brisbane, a semi-tropical area still, but not quite so stewing. Our hosts there were Robin and Robyn Black, a very welcoming couple with three sons. We took a day trip on the Brisbane River, and saw many large fruit bats hanging down from the trees. They only fly at dusk apparently just like our very small bats. We were also able to go to the city opera house for a performance. From there it was back to Sydney for our flight home after a thrilling but tiring fortnight.

Another visit to the Melbourne office was in connection with a scheme on the Weipa River in the northernmost area of Australia, the Cape York Peninsular. There was a proposal to mine there, and a water supply would be needed. This meant building a dam across the estuary to form an impoundment and my advice was needed about its spillway and the closure of the dam. This is the operation required finally to keep the tides out, normally by dumping rocks of an appropriate size progressively to narrow the gap, and this requires knowledge of their stability in the fast flows that result. In the early days in London, the hydrology consultant, with whom I share secretarial help, and with our technical staff working side by side in the office, and often collaborating in the field, was Frank Law; he was later the joint head of the Melbourne office. He had left the firm by the time it was taken over by the American

company, Black and Veatch, and found a new position as Deputy
Director at the Institute of Hydrology, the immediate neighbours
of HR Wallingford, and he and his wife live in the town, so we
see them occasionally. Again an illustration of the contacts that
those in the profession seem to keep up.

CHAPTER 14

South America

THE TWO COUNTRIES VISITED IN SOUTH AMERICA are very different, Brazil and Peru. The scheme in Peru was to upgrade an existing system that transfers water from the wet side of the Andes to the dry side. The capital city, Lima, is in a semi-arid zone not far from the Pacific, and rarely has actual rain except in those years referred to as El Niño when the weather pattern changes with clouds then coming in from the west to deposit their rainfall where they reach the Andes, when much of it falls as snow. This is referred to as orographic precipitation because it occurs where the clouds have to rise over the mountains. In other years, the weather pattern is of winds from the east which have crossed the bulk of the continent before shedding their rain – or snow – on the east side of the high Andes. In these years the only source of moisture west of the mountains is the sea mist and some of the ancient civilizations there had schemes for collecting and storing any resulting precipitation. The scheme for transferring water from one side of the mountain ridge to the other to provide water for Lima is a clever one. Water from several mountain streams fed largely by snow melt collects in a lake, the level of which had been raised by building a bank, and from there it is pumped up the steep slope to enter a tunnel dug through the mountain to emerge on the other side, flowing down the valley on the west side in which Lima is situated. There are several power stations in this valley, so the water transferred generates power on its way down, and power cables take some of this back over the mountain to drive the pumps. There is therefore an energy gain. Our job was to work up a scheme for taking water from the much larger Mantaro River, a tributary of the Amazon, pass it through a settling area to remove any sandy particles, pump the flow up to the existing storage lake half way up the mountain, provide a new pumping station there to lift the flow to the entrance into the tunnel, and so on to the other side. Crucial to this was how much flow could be passed through the tunnel, which would define the maximum size of the scheme. That was where my knowledge and experience came in, together with designing the sedimentation basin.

I and several colleagues flew to Lima, where we spent a night or two before setting off up the mountain. The Andes are very high, and the tunnel was at 4,500 m (about 13,500 feet) above sea level so over three times as high as the highest in the British Isles. Air at that altitude is rare, and mountain sickness is a worry. In fact, the week before the firm's consultant on water quality had been taken up there, but had to be brought down again hastily because of mountain sickness. We therefore had cylinders of oxygen with us that we could take a sniff of if need be, and large pills that were supposed to increase one's resistance – and some Kendal Mint Cake! We did the journey in two stages, stopping a night at a height of a bit under 3,000 m, at a small town called Matacano just the other side of a high pass (Fig. 36). Next morning we set off to complete the journey to the site, near the exit from the tunnel, where there were still the old Nissen huts from the camp where the tunnelers had stayed. That was to be where we would stay the next night.

We set off after breakfast through a fairly empty landscape, stopping briefly at a small town, La Oroya, where there was a market and all the locals were in their very characteristic dress. There were alpacas grazing looking rather superciliously at us as we passed. We were well above the treeline, of course. We arrived at site about lunch-time and all seemed well so it was decided we should press on with our work and make our inspection of the tunnel that day. The tunnel had been blasted from the rock and was unlined though some of the worst sections had some corrugated iron roof supports. It had been very difficult for the miners because of the amount of water that came in and this was flowing fast about 30 cm deep. Mostly this was under a platform of railway sleepers (Fig. 37) on which ran an electric locomotive pulling a truck with benches, a leftover from the construction, so we had a ride of perhaps a kilometre before the platform for the track came to an end and we had to walk against the force of the fast flowing water for about another kilometre before the slope of the tunnel was in the opposite direction and therefore full to the roof. It was quite tough going. We had boots on and miners' helmets with lights, as we checked on the rock quality and how rough it was – from my point of view roughness was all important as it determined the resistance to flow and hence what the capacity was.

We got back to the camp site, which was on the snowline so very cold. We had a simple meal there and I soon began to feel unwell, so took to my bunk with an oxygen bottle. I had mountain sickness and was indeed violently sick. We should not have started any physically hard work before we were acclimatised, especially me, the oldest member of the party, but I was well cared for by the youngest civil engineer of our group. I slept and gradually felt a bit better, so was able to get going next day taking things gently. This included checking on the flow measurement facilities there, now measuring the infiltration, before we went a couple of hundred metres down the mountain to inspect the Mantaro River. I was soon much better. It was different from any river I had ever seen, as it was in an area of head-sized boulders, with very little variation in size of those in the bed or banks, although the narrow plains either side had a thin covering of soil.

One of the standard measurements taken when driving a rock tunnel is the amount of 'overbreak'. This is how much extra rock has had to be blasted away to leave the required minimum cross-section everywhere, and is the basis of working out payments to the contractor. We had access to the cross-section surveys on which this was based, and there had been research in Norway on the hydraulic resistance in unlined rock tunnels in relation to overbreak. I therefore had the information I needed apart from trying to add in the resistance of the platform and its supports. Of course there was a lot of detailed work for the engineers to carry out in determining the final layout and design details of the new work. The site visit complete we returned to Lima and, as it was all downhill, it was just a very long day's drive.

Back in Lima, I was able to arrange a private excursion. My younger son, David, now an interior architect with his own business in France, had been to Lima perhaps a year before when in his first job after graduating he was working in intermediate technology, specialising in training people in building technology in the Third World, using local materials and skills. He went to many countries, with the main activity being making corrugated cement roofing sheets, using local fibres as reinforcement, with a special system for transferring the flat sheet to a corrugated mould before it set. Just such a small factory had been set up in Lima and I was able to inspect with great interest just what he had achieved. In fact when he was there, the President's English wife heard of it and he and a colleague were invited to join them in the palace for dinner one evening! I was proud of his achievement (Fig. 38).

Brazil is the only other South American country where I have been: a very big country of tropical rain forest in the Amazon basin. The work in Peru took me to the Mantaro, one of its tributaries in the foothills of the Andes and my next trip was to the River Capim, down in the rainforest, a lesser tributary yet still over 500 miles long, more on the scale of the Thames but very different in character. It joins the Amazon at Belem, a port near its mouth. The task was to advise on what river training works might be required so that large barge trains of copper ore could be brought down to Belem where they would be transferred to bulk carriers for export. The economics of the operation needed not only a knowledge of how much capital dredging would be needed but whether much maintenance would be involved and what the flow regime was, which would determine for what part of the year, if any, there would be insufficient water depth. We therefore had with us Mike Mansell-Moulin, who was the firm's hydrological consultant with an office near mine on the seventh floor of Artillery House. The only flights from the UK were to Rio de Janeiro, from where we caught a local flight to Belem, 1,500 miles on the route back home! After a night's rest, it was arranged that we took a small private plane to inspect the river from the air, to give us an idea of what it was really like. It was a very meandering river, with many oxbow lakes where earlier tight loops had cut through the narrow neck to take a short cut, just as in the geography textbooks. We could also see forest fires burning some way away, where the forest was being cleared for agriculture, basically in those days to grow pasture to graze beef cattle for the American market. However, the soil under the forest canopy is rather thin, poorly nourished and these beef farms would be abandoned after relatively few years, when the heavy rainfalls would erode the soil away to give barren areas. It was a short-sighted policy, but the local population had to live! Flying over this tortuous river in a small plane at low level meant we were banking to right then left and occasionally doing a 360-degree turn to have another look. I had to reach for the sick bag!

Next day we began our river excursion. We were to spend five days on a river boat, which was about thirty or forty feet long, with cabins in the middle, and our party, including the representatives of the mining company and an expert on barging operations, numbered five or six (Fig. 39). There was a crew to look after us of course and a crate of live chickens near the bow. It was very simple accommodation and morning ablutions were taken by dropping a bucket tied to a rope over the stern, checking the contents for piranha fish and so getting a cool wash and shave. There was a pleasant deck with an awning over it and some wicker chairs for us to lounge in as we took in the surroundings. I was busy taking photographs and making other notes, discussing problem zones such as shoals and sharp bends which might have to be eased to get barge trains round. We stopped from time to time en route at local villages, for example where there might be a river level gauge, extending perhaps twenty feet above present water level, indicating just how high the rivers in that region rise in the flood season. The villages consisted of simple huts around a clearing, with very brown people, living a simple life, fishing for their food from canoes, with small dark-coloured pigs and chickens all around. One of our pauses was where some men were surveying close to the river, guarded by a man with an old rifle. On enquiring the reason we were told that it was in case of snakes, as they had once lost a man who was swallowed by a python. No one swam in the river because of its nasty contents, not only piranha, but electric eels and alligators.

One evening, where we were to tie up for the night, we saw there was a bungalow on the bank, some fifty feet above water level with a flight of steps up to it. There was also a set of water level gauges so one of the party called at the house to ask if the resident kept the records (Fig. 40). He did, but his main job was working for a logging company. He was French and invited us

Europeans to join him for dinner. He did not have many visitors in this pretty inaccessible spot. We enjoyed his wine! Also it was a change from the chicken and tinned food we otherwise lived on whilst on board. Sleeping on the boat was not uncomfortable despite the heat – we were about fifteen degrees south of the equator – but could be quite disturbed. There were splashes, or rather the sound of water being thrashed about, especially if we were close to an oxbow. Apparently the rule for survival is that fish and alligators eat fish, so all the fish and alligators live on smaller ones, and thrash about in the competition for biting off the best bits of their prey!

On our fifth day, we got to the only point where there was a road connection and a car ferry across, just a small chain-hauled affair that could carry a small lorry or bus, or two or three cars. We were told that a month or two before a local bus had slipped off the barge through a faulty brake, but none of its passengers were drowned – they were otherwise disposed of by the piranhas, which come in a wide range of sizes apparently. It had been an amazing experience to do such a river trip, and to see the impenetrable rainforest at close quarters. Getting off the boat where one could walk across a sand bar was not unpleasant in the tropical sun but you were well aware of the hard but simple life the local people have. Our four-wheel drive vehicle was there to pick us up and drive us back to the hotel in Belem, a trip of about 100 miles, though the very tortuous river had meant that our much slower boat had taken five days to do the same journey. On the road back to Belem we passed through what could best be described as a one-horse town from the American cowboy movies, where the ranchos were strutting down the one dirt street between the single storey wooden buildings with their six-shooters at their belts, and their horses tied to the hitching posts.

Next morning I made an early start to the airport to begin the journey home, going generally in the wrong direction first, back to Rio. I got there mid-morning but my flight to the UK was not until the evening, so what was I to do in the meantime? I had seen nothing of the city when I first arrived because we had a fairly direct connection to Belem. I saw there was a tourist desk in the airport arrival area so enquired if it was possible to do a city tour to fill in my time. Yes, it was and it would leave a point in the city at 2 p.m. and return in good time for me to catch the bus back to the airport. I bought my ticket, got my bus to town, bought a snack, and wandered down to the beach area where I sat on the sand, my back to the wall of the promenade, feeling very incongruous in my suit amongst the folk in their beach clothes or swimsuits. At 2 p.m. I was back to pick up the tour bus, and so off we went. There was a chap took the seat immediately behind me and we got into conversation: he was also a Brit. but rather than flying home that night he had another day to spend in the city. He was a hydrology consultant working for one of Binnies' rivals in connection with a water supply project. We had a good tour of the city and its surroundings including going up the peak to the huge figure of Christ there, with a view across to Sugar Loaf Mountain and of the bay area.

However, there were storm clouds around us, a tropical storm was brewing up and the heavens opened on our drive back to the city. We were using a major road close to the bay but the drains could not take away the volume of water, which formed pools up to a foot deep in any dips in the road. Many cars had stalled and tended to block the road so we were making only fitful progress. Time was getting on and I had visions of missing the airport bus I had to get to catch my flight. Perhaps best part of an hour had been lost stuck in traffic and when I told the tour leader of my problem she spoke to the driver, who fortunately knew that there was an alternative bus to the airport that was due to leave a hotel not far from our route in ten minutes or so. He was able to get off the main road and into the hotel forecourt just in time for me to catch it – and I heaved a sigh of relief. I had left my suitcase in a locker at the airport so I caught the flight home after an eventful day. The flights to and from Rio de Janeiro in those days were by VC 10 aircraft, a fine but noisy British aeroplane with four engines near the tail.

Go forward about five or six years, back in Moulsford and four doors away one of the bungalows in Glebe Close changed hands. Our immediate neighbours decided it would be nice to welcome the newcomers to our group of good neighbours by holding a drinks party one evening. The name of the newcomers was Parker and the husband was much my age, and nicknamed 'Pickles'. He seemed vaguely familiar and when I told him I was a civil engineer and that I worked for Binnies, he said he remembered meeting a Binnies' man on a tour bus in Rio. Extraordinarily we had met before! It's a small world, but sadly both he and his wife have died, and the property has changed hands again.

CHAPTER 15

UK projects, university and committee work

SEVERAL OF THE UK PROJECTS I WAS involved with have already been mentioned, including the work on the tidal power project in the Severn estuary, experiments in the Wash on slope protection and research into the resistance of tunnels lined with precast concrete segments (Fig. 42). However a lot of my time was taken up by the day-to-day activities of a consultant in a large firm – and Binnies had a staff of the order of a thousand. I always adopted an open door policy (this was before the days when there were large open-plan offices). If anyone had a hydraulics design problem which they did not quite know how to resolve, they would knock on my door and ask if I was free to talk about it, or else ring me and ask if they could come. Usually I could refer them to a chapter in a text book or design manual, or perhaps I would need to look up a paper that would be useful for them in the large card index I kept of published research. I referred to this as a grasshopper mind, ready to jump from one topic to another at very short notice. Also I worked on Binnies' own hydraulics manual, intended to cover most standard theory and to be of use in offices overseas or on site. It was also my job to supervise any model studies being carried out but we also had our own small laboratory in another building about half a mile away where the water quality lab was, so we could do some small models of our own.

Of course many queries came from overseas offices and the largest branch was in Hong Kong. In the seventies the method of quick communication was the telex, basically the same system used by the post office to send telegrams. There was a telex-operating girl, who punched out paper strips in a code of holes a bit like Braille, and then these were usually sent overnight. This provided a pretty quick exchange, but the London partner most involved with the work in Hong Kong, Stanley Ford, had a loudspeaker

conference phone so that we could assemble an appropriate group
at both ends, when times were compatible, to discuss major
problems or proposals and what they might entail. Later we were
able to update to fax machines, and one useful aspect of this
arrangement was that the offices in Melbourne and Hong Kong,
with big time differences, could get a turnaround of under
twenty-four hours. As long as they got a fax to us by the end of
their working day, they would get the answer when the office
there opened next morning, as we always got a reply on the way
to them during the same day that the query arrived in London.
Now, of course, emails have taken over from the fax machines.
Part of my time was taken up in writing reports responding to
queries or writing up reports on work overseas when I got back to
the office. The usual practice was to write longhand and pass it to
my secretary, Carol, for typing, but more commonly with
day-to-day correspondence I could dictate to her.

So much for the everyday routine but there were projects
requiring field trips; for example I and several colleagues inspected
the proposed site for the Severn Barrage by helicopter, and there
were many other field trips to project sites. My team of
well-qualified young engineers were themselves good at hydraulics,
of course, so they were usually directly concerned with the design
of many projects the firm was involved with, sometimes going
overseas to carry out surveys or work on schemes in overseas
project offices alongside local staff.

The work on the possibility of building a further London airport
on land to be reclaimed from the sea at Maplin Sands off the Essex
coast has already been mentioned in that it led us to develop a new
method for the probability of high surge-affected sea levels, but
there was more to it than that. We were also assessing wave climate
and how the waves would be refracted over the shallow sea. This
process is a bit like the refraction of light through a lens and it can
similarly concentrate the waves in some areas or weaken them in
others. There was a computer programme for doing this and I was
very lucky to have a team of brilliant and energetic young
engineers to work on the details, and in view of the short time scale
we were given, we had meetings every two days to discuss
progress. Binnies were also involved with proposals to build

reservoirs in the sea by constructing bunds of dredged sand to enclose an area into which fresh water would be pumped. The proposal was to pump some of the dredged material into large sacks, to form front and back walls of the bund, the exposed side then having to be protected against the wave action by a layer of rock. This scheme was in the Wash, on either side of the estuary of the Ouse. When we visited that site, we would get the evening train to King's Lynn with a good meal en route and then would go out to site, using the contractor's hovercraft when work on a trial bank got under way, but otherwise walking out from the nearby coastal bank over the mud and sand, with a wary eye on tide times. Getting through the mud areas was tricky, as it was so glutinous and in some places almost a foot deep – a very tiring exercise. There was a nice hotel in the town square which specialised in samphire as a starter, a strange salt marsh plant with fleshy branches around a stringy core. One of the ideas being tried was the use of tetrahedron-shaped bags with 2 m sides, easy to sew up from the plastic sheet material, but not a success because they could not be stacked effectively.

Another of Binnies' jobs around that time was a scheme to clean up a river in Staffordshire (the Tame) by diverting it through lakes where polluted sediment, some no doubt coming from sewage treatment plants upstream, would be settled out and then separately treated and disposed of. These ponds had inlet and outlet arrangements needing hydraulic design (Fig. 41). Another job stemming from the dirty water side of Binnie & Partners' business was a scheme for bringing together the sludge produced at sewage works at several Lancashire towns so that it could then be handled more efficiently. This meant long lengths of pumping main and the head loss in these would determine what pumps would be needed. Because of the glutinous nature of this sludge, centrifugal pumps are not ideal so ram pumps are used, having pistons to push the sludge down the line. Every town's sludge is a bit different and there was no reliable method of design because the sludge is somewhere between a true fluid and a paste. It was necessary to test the sludge from each town in a rig, involving a loop of about 10 cm diameter pipe, and a smaller rig to determine the rheological properties. A local Binnie engineer carried out the actual testing, not at all a pleasant job, but after designing the set-up, I was able to show him how to analyse the data so that it could be utilised in scheme design. I was back to being involved with sewage again, as in my local authority days!

Subsequent events made one incident stick in my mind and that was being asked by the engineer designing a water transfer project for advice on the ventilation of a tunnel. This would connect from the River Lune in Lancashire to the River Wyre, so that water could be transferred between them in case of a water shortage affecting intakes on the Wyre. He brought the design at the Abbeystead end of the scheme for me to see, and his specific query was what area of ventilator was needed when the tunnel refilled from the far end after any period of shut down. This was reminiscent of the sewage pumping stations of my early career in that the tunnel formed a wet sump for the equipment in the building above. One principle of design firmly in my mind was that there must be no connection between the wet well into which the flow came and the dry well in which the electric motors and similar equipment were situated. This was because of the danger from sewer gases, not just because they might have asphyxiating elements but because they might explode, as septic conditions in the sewers could generate something akin to the firedamp in coalmines. However, this was a building in a hillside and therefore partly underground, with steel sliding doors between the tunnel (the wet well) and the equipment chamber. I said I did not like it at all. It was explained to me that the operating rules for the station would ensure that these steel doors were firmly closed after any shut down, which was why there had to be some other route by which the air being pushed ahead by the water filling the tunnel could escape, and environmental reasons stopped them providing a separate vent shaft. The vent would therefore have to be in the upper wall of the wet well. I explained that this could simply be worked out by setting a maximum reasonable velocity – but that despite his assurance over operating rules, I still did not like the proposed arrangements. I argued that there might be algae or similar deposits in the tunnel even though it was carrying river water not sewage. I nevertheless remained reluctant to accept his assurance that there really was absolutely no risk as river water was very different from sewage, and I said so in no uncertain terms, but I was glad I had no responsibility for the overall concept or details of design. In due course, the scheme was built.

I decided to take partial retirement when I was sixty (1984), something I had always had in mind because standard retirement age in the public sector had been sixty. This was also about the time that Binnies decided they could no longer keep pace with the cost of leasing property close to Victoria Street: the lease on Artillery House was due for renewal and would cost a lot more. The decision was taken to move to the southern outskirts at Redhill, where rents were much lower and there was a new office block on the market not too far from the station. However, it was the wrong side of London for me to commute and we were so happy in Moulsford that it made sense to stick to the decision to scale down activities when I was sixty. There were still jobs at Binnies that might involve me but I could handle these by ad hoc visits, travelling by car. Also I could take on requests from other firms to do specific jobs for them from time to time, but I could control how much professional work I chose to do.

It was a year or two after this that on the evening news one day there was a first report of a major explosion in Lancashire. The next day's news brought the horror of the tragedy into the headlines. It was indeed the Abbeystead valve house, and sixteen people had lost their lives. Another twenty-two were severely injured. What on earth had happened to cause an explosion with such a huge loss of life? It emerged in due course that it was a gas explosion, which had caused severe damage to the building, whilst members of a local authority committee were visiting it for a demonstration. It was indeed a catastrophe with an almost unbelievable number of casualties. It emerged that the committee had been invited by the authority operating the water transfer scheme to observe it coming into operation after a lengthy period when it was not needed. What had happened to those operating rules which should have prevented such a calamity happening? The tunnel was supposed to have been kept full of water. The steel doors should not have been open when the scheme was in operation, but they were wide open so that the visitors could see what was going on. Whether it was a cigarette that ignited the gas which was pushed out of the tunnel or an electrical spark was probably not known: there was no smoking ban there. Those affected by the tragedy, especially the dependants of those killed including many children, deserved compensation and the parties concerned were properly insured. However the authority's and engineers' insurance each denied liability so it became necessary for them to take to the courts to make one or other of the insurance companies pay up.

In cases like this it is not the insurers who are the main players in court but the parties they insured, so it appeared as if the water authority was suing the designers, Binnies, for professional negligence, whilst behind the scenes the insurance companies would be employing the legal teams and keeping things very much under their control. In the event, the case went largely against Binnies, the other side having brought in an expert witness to testify that they should have been aware that there were coal measures under the site and that the methane gas trapped there could ooze out and get through the concrete lining of the tunnel into the empty space, where it would stay because it is heavier than air. What Binnies' defence was, I don't know but it then went to appeal, when three Law Lords reviewed the case. Number one said the verdict should stand; number two put a well-reasoned argument that Binnies should not bear the brunt of the blame, it was an accident due to unforeseen circumstances; number three, now with the unenviable duty of the deciding vote, ruled that the arguments were irrelevant; what mattered was the people affected had to get their compensation and if he found for Binnies they would not. I am sure it was the right decision, but what a shame that the families of those killed had to go through all that trouble to get what they clearly deserved. I don't think it affected Binnies' reputation very much because most engineers thought 'But for the grace of God there go I'. I was not involved in the case, to my great relief, but was my objection to the basic design some sort of premonition? I think not – and I certainly had not based it on any knowledge of the coal measures. It does not matter where the true blame lay, but it shows how tragic accidents usually stem not from one unexpected occurrence but from a combination of them. In this case the unexpected elements, the absence of any one of which would have avoided the tragedy, were: the tunnel had been kept empty rather than full; the steel doors between it and the equipment area were open; there was a large number of visitors in the pump room at the time; and perhaps also the absence of a smoking ban. I felt sorry for my ex-colleagues who had, in effect, been accused of professional incompetence at a late stage of their careers, but these were conscientious and competent engineers. Was the basic design of the project mostly to blame, or was it a serious mistake in the operation by the authority staff?

Working in London, which I did for some twelve years, meant a daily commute. My neighbour, Sandy Thomson, who worked at the Air Ministry, usually shared transport to the station, and in theory I could get to Artillery House, via Paddington, in an hour and a half. In reality, it was often longer because of problems on the line: frozen points, signal failure and other trains breaking down were frequent excuses. It was a choice of Circle Line or the number 36 bus to get to Victoria. The Circle Line was unreliable because it seemed as if the train drivers or guards, who had been at work for a couple of hours before the time we got off the train, took their first tea-break of the day just when the main flood of office workers arrived, so there were gaps of twenty minutes in the service, and often trains were just cancelled. The number 36 was OK until new front-loading buses came into use, when the practice seemed to be to run them in convoys of four or five. As one waited for the convoy to arrive, the queue getting longer all the time, the first bus would draw up: 'First three only, please!' Meanwhile the second bus driver with his door nearer the back of the queue would fill available spaces with passengers from the middle, and bus number three would have taken anyone quick enough to get in from the very back of the queue before he closed the door and drove past the rest of us with plenty of room. Bus number four was empty so didn't need to stop to set down passengers so overtook the buses in front – and there the early people in the queue were left to wait for the next convoy!

Commuting was never pleasant though the train journey did give opportunity for the technical reading I needed to do to keep up with research – though coming home at night there was a fair chance I would drop off to sleep. It became routine but my neighbour who got to Paddington just in time to get on the usual train did not notice that there had been a platform alteration. The train did not stop at Reading, where the normal train had its first stop. Why was this? He was on an express to the West Country which did not stop until it got to Taunton, about four hours away! When he got there, he found the last train back had already gone but the railway staff arranged for him to travel in the guard's van of a late night mail train, which dropped him at Reading in the early hours, where his wife was able to pick him up.

The worst experience of my commuting days was, however, being involved in the rail crash near Ealing. We were at top speed when there was a sudden jarring and we realised the coach was off the track and was juddering over the sleepers until it came to a rapid halt. We had indeed been derailed! We were unhurt, only shaken, but realised the danger if another train ran into us as we were in the rear coach, so it was a case of picking up one's belongings and heading for the door. There was a railway employee there first, head out of the window, shouting 'Guard, protect the train!' The guard's van was equipped with a short-circuiting cable with which he ran to the back of the train to put it across the two rails, when the short-circuit would set all signals to red. Actually it transpired that the front half dozen coaches had fallen over and spewed across several tracks and indeed all signals were already at danger. We jumped down to track level in turn (trains are surprisingly high), and found someone had put a ladder over the back fence of the gardens behind a row of terraced houses. We walked through the house at the lady occupant's invitation, meeting firemen going through in the opposite direction, less than ten minutes after the crash. Their station was close by and they had heard the sound of the crash. I was then in a virtually empty main road and my first thought was to ring up home to tell Margaret I was safe and would try to get home somehow. I found a telephone and realised my best move was to go towards London to Ealing Broadway, which had a surface local line to the City, so was able to get the bus from Victoria that in those days ran a service to Wallingford. Another telephone call home meant that David, who could then drive, could meet me off the bus.

It was this event that made me appreciate how efficient our emergency services are in such a disaster: it had indeed initiated the disaster proceedings. The roads in the area were closed to all except emergency services: doctors were being brought in from local hospitals with a police escort; ambulances seemingly by the dozen were arriving; firemen with heavy lifting gear were on their way. We in the rear two coaches were the lucky ones, our coaches having stayed upright. The locomotive was on its side several tracks away, and between it and the upright coaches were the front coaches, not merely on their sides but some had crashed over the top of others. Over ten passengers were dead, and very many more injured, some very seriously, and we lucky ones had escaped without really taking in what horrors there were near the front of the train. This was brought home to me a few weeks later, the track by then having been repaired, when the group in the compartment leaving Cholsey started talking about the crash and I said how lucky we all were to escape. One of the City gentlemen there said 'Speak for yourself. I shall never forget that night, as the wheels of a coach riding over us cut through the side of the carriage and decapitated one of the men in my compartment.' I was indeed so very lucky on that occasion to have chosen to sit at the rear of the train.

Perhaps the largest and certainly one of the most challenging and rewarding of the jobs in the UK was the study of the Severn tidal power project. Here Binnies were the lead consultants to the government committee that was set up to examine the project in detail, though we had the support of other consultants dealing with construction methods, the generation and transmission of the power and any environmental effects. We needed a sizeable team and the firm allocated its best brains to the project. The other two engineers who joined me to plan the investigation were Clive Baker and David Ruxton, and we had several meetings during which we prepared a chart showing how the study would progress, month by month, with all the necessary interactions, through the study period of eighteen months, not a long time for such an important study. There were many aspects we could do in-house, for example, a first numerical model to consider a range of possible sites from English Stones out towards Minehead. This model would simulate the full range of tides, from springs to neaps, take account of the way in which, as the tide rose, gates would open to admit flow upstream past the barrage, then when the tide turned, another set of gates would open to let the flow out through a set of bulb turbines (axial flow turbines and their generators, set in a bulb in a large tubular water passage). The hydraulics of the turbine and gate passages were included, as well as the anticipated turbine characteristics, i.e. the relationship of the electricity output to the head through the turbine, basically the difference between upstream and downstream water levels. Of the lines examined, the one between Brean Down in Somerset and Lavernock Point in Glamorganshire, going through the islands of Flat Holme and Steep Holme, was found to be the most effective in terms of the amount of energy developed compared with the probable cost of construction. This was a line that Professor Eric Wilson, a strong supporter of tidal power, had suggested would be the optimum (Fig. 43).

The method of construction envisaged was by using caissons to house the sluice gates and turbines, which would be towed out and sunk in place during a slack part of the tide, rather like the caissons for Mulberry Harbour had been floated across the Channel and sunk to form a prefabricated port for the British troops in France in World War II (Fig. 44). There was field work to carry out to measure currents and waves, because the remainder of the gap not occupied by the caissons would be formed of rock and earth embankments, which would need protecting. Also a two-dimensional model would be needed to show all the current patterns that could not be shown in the in-house one-dimensional model, and this was allocated to HR Wallingford. Also there was concern for whether the scheme would 'de-tune' the estuary so that tides would diminish, and this involved another 2-D model going further out into the Irish Sea, carried out by the Institute of Oceanographic Sciences in Bidston. The scheme would obviously affect shipping, which would have to go through locks being proposed towards the Welsh side, so the modelling also had to show how much time would be available each tide for the ships, including very large ones. to continue accessing the ports upstream such as Avonmouth and Cardiff. What would be the effect on fresh water floods? This was resolved by using the 1-D model, inputting a flood, and showed that the flood levels upstream would reduce, a factor that now has added significance as sea levels rise due to global warming.

The study included such items as access for all the materials required, employment and the effect on the environment. Salinity was covered by the in-house model, and there would be a slight reduction upstream. Sedimentation was an important issue, as the Severn is known to be very heavily sedimented, confirmed by our field work, with a peak during spring tides and a considerable drop during neaps. This occurs because it is a closed system: in other words, the sediment there does not come in from upstream (or downstream) but is always there, being picked up from the muddy bed during spring tides and deposited again in the gentler neap tides. Knowing from the model how currents would vary, and in particular be greatly reduced upstream, meant that we were able to predict with some certainty that the estuary would become much cleaner, with water much more transparent, so that the extra light would allow increased biological activity. Also there would never be a time during construction when the estuary was closed off to tides, as the tidal flow would continue to ebb and flow using the openings in the caissons, which would only start controlling the flow when the scheme was complete and ready to generate power. This would go through a transformer station, and via new overhead cables to the grid. CEGB, the Central Electricity Generating Board that the country then had, used its own model of the whole system to show that the output from the scheme, about seven times the output of our largest nuclear plant, could indeed be accepted: there was no need for dedicated storage.

All subsequent studies have confirmed our findings, which we put before the committee under the Chairmanship of Sir Herman Bondi, as they arrived. Of course a number of different hands were involved in drafting the final report of the committee as it covered so many different specialisations, but I did much of the sections relating to hydraulic matters. The committee sat under the umbrella of the Energy Technology Support Unit, based at Harwell, and they virtually had the final say in what went to government, and one of the issues that people had in mind then was what the electricity would cost in relation to nuclear power. It was indeed competitive, even though the estimated costs had a 10 per cent addition applied to them in case of cost overruns, and the nuclear costs did not include any costs of waste treatment, storage or disposal, or decommissioning. (The next phase of nuclear development included Sizewell B, which was many years late in delivery with large cost overruns!) The report sat on the Minister's desk for some time when, suddenly, the then Thatcher government resigned and all unfinished business was rushed through on the last day of Parliament, virtually without discussion, including the Severn Barrage committee's report. It was wrong to see tidal power as a direct competitor to nuclear power, but as the government's chief scientific adviser at the time was Lord Marshall, ex-director of the Atomic Energy Research Establishment, it is not surprising that the Prime Minister took his advice. After all, it had been the base load of atomic power that had enabled the country to keep going on a three-day working week during the miner's strike.

I think we might just have seen the process repeated, as, despite its many advantages, the Severn tidal power project is unlikely to be chosen as a means of reaching the present government's target for renewable energy. This is a shame. It would have an installed capacity of seven times Sizewell B, our largest nuclear station; it would have a life of at least 100 years, long past the expected life of any alternative; it would protect Gloucester from the worst effects of flooding when there are high tides, and against rising sea levels; it would produce much more benign conditions upstream because of the reduction in tidal currents, so that pleasure boating could thrive; it would bring the sea closer towards resorts like Weston-super-Mare because the tide would not go out so far; it would produce electricity at a very competitive price; there would be no hazardous wastes to store until our grandchildren and great-grandchildren discover how to dispose of them safely; only hydropower and wind power use essentially free fuel, and, like coal, this does not need importing whereas uranium ore, oil and gas all put the nation at the mercy of unscrupulous foreign owners; and it would improve water quality so much that the estuary would be much more ecologically active.

I think it was getting to know me through work for that committee that led to Sir Herman Bondi subsequently inviting me to become a member of the Natural Environment Research Council (NERC). I knew something about tides and floods, but not much else that was part of the natural environment, and was covered by the Council's many national research laboratories, including the Institute of Hydrology next door to my old stamping ground in Wallingford, the Institute for Geological Sciences, the Institute for Terrestrial Ecology, the Institute for Oceanographic Sciences and the Liverpool Tidal Institute. Membership of the Council meant that I learned quite a bit about these other sciences, surrounded by academics, each an expert in their field. In fact, after a spell as a member of Council, I was proposed as Chairman of the Marine Sciences Committee. Perhaps it was thought that someone like me would bring a more down-to-earth approach when we were discussing how to spread out the funds! I did not have an axe of my own to grind. It also meant some rather special trips, some including Margaret, such as the launch (or rather commissioning) of the then new IOS research vessel at Barnstaple, a meeting in Cambridge and dinner with the Master, preceded by aperitifs at his home.

Part of the duty of the Council of NERC was to carry out surveys of the science taking place at each laboratory in turn, a sort of science audit, several of which I took part in, and so saw work at the frontiers of those sciences. When we visited the Institute of Hydrology for this purpose, we also visited the Culham Laboratory, when it was nearing completion. It was to research into fusion, in the hope that ultimately fusion could take the place of fission-based power installations. It was quite impressive to see the level of sophistication in the equipment being installed. For example, because of the intense magnetic field that would be generated, all instrumentation had to be based on glass-fibre technology, as wires carrying electric signals would not work. However, although even bigger and better establishments are being set up overseas to try to find a way of containing the extremely hot plasma, in my view the chance of ever producing electricity at economic cost is minimal: after all, it is just another way to boil water and produce steam, in order to drive turbines connected to generators. The fuel may in theory be cheap, but the capital cost of providing a new type of fusion boiler would surely rule it out – and despite spending a lot of money on 'matches', there has yet to be any sign of the 'ignition' of a sustainable 'fire'! Research continues of course, and there may yet be a breakthrough.

Other committee work has gone on through much of my career, from the Rainfall and Run-off Committee from around 1960, a committee to study the effects of marine outfalls for sewage which published its findings with the unfortunate title 'Out of sight: out of mind'; the Hydraulics and Public Health Committee of CIRIA, the construction industry's research body; being on the Council of the British Hydromechanics Research Association, and later on the Board of the Hydraulics Research Station in Wallingford; also the British Standards committee on Flow Measurement by Weirs and Flumes, and its international counterpart. There was the British committee of IAHR, the International Association for Hydraulic Research, which led in turn to my appointment to the Council of IAHR. This was indeed a prestigious appointment because it was a recognition of my international standing in hydraulics at that time, 1970 to 1974. That was the maximum term one could serve on the council unless one became a vice-president prior to a term as president. This meant attending the annual council meetings, every other one coinciding with one of the association's big biennial conferences. In fact during the 1970s I attended most of those conferences and Binnies kindly supported me in doing so. Also our children accompanied Margaret and me on two of these occasions, John and Sheila being with us at the Paris conference and David at the Istanbul one. There were programmes for the accompanying persons that they could take part in whilst I and my colleagues on the council were slaving away at our meetings!

Actually it was far from being all hard work and the conference in France, when the council members were the guests of the French government, was especially lavish. We arrived in Paris in time to join the train for Tours in the Loire Valley, heading ultimately for Marcay, where our meetings were to be held and where we also stayed. A super meal had been laid on for us in the train – which broke down about an hour out of Paris! So it is not just trains in the UK that are prone to problems. We were being well-refreshed of course, but it meant that we arrived at Tours quite late, and there was still a bus ride to get us to our destination. There was a lavish buffet supper laid on for us, but really we were too tired by then to take it all in, having had an early start from London, and done a little tour of Paris on our own before going to the station to meet up with our fellow attendees. Arrangements had been made for John and Sheila to stay at a coach house, not far away, so they were also well looked after, while we stayed at the government-owned chateau. Highlights of the stay included an alfresco party, with an absolute mountain of beautiful food – and the French know how to do that better than anyone else. There was wild boar roasting on a spit, and the best champagne and wines. Then we heard huntsmen's horns in the distance, coming closer and closer whilst playing a distinctive air. They came into sight in all their finery – and then we turned our attention to the display of sweets!

The meetings of the council took about five days, and during this period the wives and our two children, then aged twenty and seventeen, visited the several chateaux in the region, Chenonceau and Langeais, and the castle at Chinon. The abbey at Fontevrault was part of the itinerary, when the council members could also accompany them on a half day's break in proceedings, where we saw the tombs of Richard the Lionheart, and the wife of King John. The last evening there we were taken to Le Lude, to see a *son et lumière* of the history of the region, much of it shared with us of course, where we sat in the reserved seats in the stand, whilst the performance took place on the other side of a small river. This was well done with dancers appearing from around the side of the chateau at the opportune time, more huntsmen, knights in armour etc. Another very late night! Back in Paris for the congress itself, we were brought down to earth, staying in university accommodation some distance from where the meetings were held. Not much needs to be said about the meetings themselves, but we did have a cocktail party one night at the first floor level of the Eiffel Tower, another rather posh occasion.

As was usual, there was a post-congress tour that we took, travelling by train overnight to Grenoble to pick up our coach in order to continue our tour. Our first stay was at a nice hotel called *Les Trois Roses* but there were some long journeys between the points of interest we visited. There was a nuclear establishment that the driver had great difficulty finding and the people there did not seem to be expecting us anyway; there was a high concrete arch dam in the Alps, when the ladies, plus John and Sheila went to the abbey at Chartreuse and sampled their liquor. We saw the many power stations on the Rhône, and also visited a major hydraulics research lab at Grenoble. We visited Avignon late in the day, seeing the *pont* that ends in mid-stream, but no time to visit the pope's castle there. At Arles, we had a very close encounter with the Roman remains as the hotel was built onto the side of them, and Roman stonework formed part of our bedroom wall! There was also the amphitheatre there to see before going down into the Camargue, a marshy area not far from Marseilles where we had lunch including eels, before beginning the journey home.

It is not feasible to detail all the conferences we went to but they provided excellent opportunities for world travel: to Istanbul, Ottowa, Baden Baden, Lausanne, Sardinia, Madrid, Melbourne and then the 26th Congress, the second one to be held in London, which I helped to organise, in 1995. That was my last IAHR conference. My services to IAHR and to hydraulics were recognised by the award of honorary life membership in 1985, perhaps my most treasured award since the award of the Royal Scholarship to Imperial College which started my career. There were other conferences not organised by IAHR, for example Gothenburg, Hong Kong (with the ex-Imperial College, Professor Joseph Lee, as an organiser) and Fort Collins in America, including a week spent up in the Rockies at their summer campus, where Margaret was able to join me. The Ottowa Congress had also provided an opportunity to visit some of her family in Guelph, Ontario, where we also spent a very good holiday at a timber-cabin hotel at the side of Lake Katamawigamog, where we were able to take the hotel's Canadian canoe out, and also visit what, during winter, would be the ski slopes and trails, a wonderfully relaxing trip. The Melbourne Congress enabled us to tack on to it a trip to New Zealand, where we toured the North Island, visiting my relations in Rotorua with its thermal pools, and also a long-standing family friend from my early childhood in Bootle, Millicent Dibble. That again was a notable experience, New Zealand being such a beautiful country, with so much to see. Our final day there was at the coast, sat at the top of a hill next to the Pacific on the Coramandel Peninsular, with the Shoe and Slipper islands just offshore, the estuary of the local river there below us, with mountains forming the backdrop, perhaps the most beautiful place we have ever been to. We were reluctant to leave, but we had to get back to Auckland and our flight home!

It was in 1973, soon after I left the Hydraulics Research Station, that I was invited to become a visiting professor at my old college, City and Guilds. Binnies were happy to allow me to spend up to one day a week during term-time on my duties there, which were mainly to help run the MSc course in hydraulics, bringing some outside experience in to temper the more academic approach of the permanent staff. The head of department was Professor Patrick Holmes, an expert on coasts and waves, and other members of the permanent academic staff were Dave Hardwick and Paul Minton, who had been there when John studied for his degree there. (We still see Paul and his artist wife because of our patronage of the Henley Symphony Orchestra, where my old colleague Rodney White plays the bass trombone!) I was still a visiting professor up to about 1983, when there was no longer sufficient support for a masters degree in hydraulics, and my duties had virtually vanished, and it was time to give up that activity. There were still people in Binnies who liked to give me the title of professor, but I was not happy to be so called outside the university walls.

Other work for universities, besides the occasional lecture, was as an external examiner, often for a PhD candidate, but frequently for a whole class doing a masters degree, as at Newcastle and at Southampton. So there was I, having had only two years at university myself getting my degree when I had just turned twenty, now examining those who had written a thesis for their doctorate! I took this all quite seriously, of course: these candidates had given several years up to their higher level degree work, and as I read through their theses I would put in some flags marking places where I thought there was some aspect I could question them on. The difficult ones were the marginal cases, and some of those were on some award from a foreign country. How could they be sent home with nothing to show for their efforts? Usually there was a way round this by setting them some adjustments to make before the examining board, which would automatically include their supervisor, would feel able to pass them.

After I had really meant to be fully retired, I was asked to be an Assessor at a couple of public inquiries, under the planning regulations. One of these was for a scheme to protect Bideford from tidal flooding. The scheme before the Planning Inspector, with me sitting beside him, was for a wall alongside the quay, perhaps a maximum height of 1 m with gaps in it to give access for the fishing boats that still plied their trade, and for the paddle steamer that still went over to Lundy Isle, indeed the same one we had used for a day trip there during our honeymoon, some forty years before. The gaps in the wall would be filled by sliding gates that would be closed whenever there was a risk of a very high tide, possibly surge affected. Obviously there were several interests that would be affected by the project, and they had put in their objections, hence the need for a public inquiry, with both sides being represented by lawyers, although there was every encouragement to keep it relatively informal so that private people raising objections could have their say. The Inspector and I stayed in Barnstaple and Margaret joined me there for the final few days, including our final site visit when the rules state that there must be no contact between the Inspector and the opposing parties. The Inspector wrote his report – he was the prime mover being an expert on planning law – and I added in my bit about the need for the scheme, its frequency of operation and matters relating to hydraulics. The report then went to the minister's desk with our recommendation that planning permission should be granted. The scheme was built and operated successfully. Margaret and I finished that trip with a visit to Ilfracombe, where we had spent our honeymoon, and we found that the hotel was still there and we could even identify the room we had, though in the intervening years Ilfracombe had gone somewhat down-market.

I had always said that I would retire completely from any professional work when I was seventy, because I reckoned that even if the mind was still operational, one would be out of date – and it would show! Having progressively scaled down since leaving the full-time employ of Binnies at sixty, I was getting close to total retirement when a call came from HR Wallingford, to help them out with understanding the results being obtained from the large-scale research then being carried out on two-stage channels. This is where a central deep section is flanked by shallower flows over berms set at a somewhat higher level. This is, in effect, like a river with flood plains, although the first stage of the research that had been carried out was a simplified case, with a straight trapezoidal channel with symmetrical berms. The measurements in the large-scale research set-up had been taken very accurately, but those carrying out the research could not really understand what they meant. Could I look at them with an independent eye and take them forward? I agreed to do this and, after another spell in the bath to help the thought processes, I knew how to proceed. Working with one of the scientific staff at HR, we set up the necessary system of analysis. By that time I had my own desktop computer, one of the first cheap home machines, an Amstrad, and this had a version of BASIC as its preferred programming language, whilst the big machines at Wallingford were using FORTRAN, a very common scientific language. We both had our independent programs, therefore, and so could check each other's calculations. The problem really boiled down to the question whether, when working out the hydraulics of such compound channels, one should treat the section as a unit, or work out the hydraulics of the deep channel and the shallower berm flows separately and then add them up. Of course it was not so simple as just making the choice: neither properly represented the careful measurements. These were accurate to 0.5% so, if one's theory was correct, should it not agree with them to something like that accuracy? After a lot of detailed work, we got there and the theory put forward did achieve the required standard, leading to several papers on the resistance of straight compound channels. Real rivers are much more complex however, as they are seldom straight and may wander from side to side of a very variable flood plain.

My last job was exactly that: a very complex example of an artificial channel specifically designed to resemble a complex natural channel, with the main channel varying in cross-section, flanked by berms with embayments that would be planted with reed beds, so producing a realistic representation of a real river when it had all matured. This was again as an Assessor at a public inquiry, this time into the flood relief scheme for Maidenhead, Windsor and Eton, in the form of a second river to relieve the flow down the Thames. The flow would be divided just upstream of Maidenhead by a gated intake to this new channel, which was designed to take 215 cumecs (cubic metres or tonnes of water per second) during a very big flood totalling 515 cumecs, the flood expected to be exceeded once in sixty years. This was a planning inquiry, not an engineering inquiry, but my role was nevertheless important as although the government's planning inspector allocated to this task knew everything about the subtleties of planning law, he knew little about engineering aspects, and several of the objectors were doing so on engineering grounds, such as that there was no need for the scheme, or that there were other schemes that might work better, or that it would make flooding downstream from the scheme much worse. It was a very challenging and interesting inquiry, with QCs representing the main protagonists and yet with ample opportunity for the private person to also have a say. The promoters of the project were the Environment Agency, which has the responsibility for flood protection in the country, and they reckoned that the scheme would need to avoid flooding in the area if there was a repeat of the flood of 1947, the worst in living memory. Flooding on that scale would indeed have a more severe impact, not only because of the greater number of properties affected but also because of the vast increase in motor traffic, and commuting by car and rail from those heavily populated areas west of London, because all road and rail links would be shut down for several days.

It quickly became obvious that the designers did not appreciate just what effect the environmental enhancements would have on the capacity of the scheme. The research at Wallingford had not yet looked at such a complex channel system and yet I knew from the work I had done on straight channels and initial results for meandering channels that it would not carry anything like the flows the designers thought it would. It would require alteration. Fortunately this was perfectly possible because the scheme contained several weirs: there was plenty of gradient available. It could therefore be steepened and the consulting engineers were asked to run their computer model with much increased roughness factors so that I could advise on what should be done in the way of increasing bank levels. The inspector's lengthy report therefore had an appendix giving my conclusions and recommendations – and indeed warnings for the future operation because there were already a couple of examples of schemes of this nature (but on a much smaller scale) failing to carry the flows their designers had anticipated. The Maidenhead, Windsor and Eton scheme was probably the biggest flood project carried out since the war, and there was no room for any uncertainty, every precaution had to be taken to see that it worked and did what it was supposed to.

The scheme was under construction for a couple of years and we watched its progress as we travelled along the M4. It was virtually complete, apart from various planting, when, in January 2003 the Jubilee River as it is now called was opened for the first time to help tackle the largest flood for very many years. Though only carrying about 144 cumecs, two-thirds of its design capacity, it started overflowing its banks and some flooding occurred that the local residents were convinced was due to this: it was a total failure, many people said. Actually it would be more accurate to say it was of limited success, because it certainly did supplement the flow capacity of the main River Thames, so probably avoiding the flooding of many houses. However, clearly something was wrong, and it soon became clear that there were sections of bank not built up to the correct level, or poorly constructed so that they were not watertight. I am not party to any information that would enable me to pass any comment on exactly what went wrong, but this was a scheme where the environmental 'enhancements' clearly carried with them very severe hydraulic penalties. Did the designers not modify their design to take account of my recommendations, hence the deficiency in capacity? In conclusion it is perhaps worth quoting some of my criticisms and recommendations as given in the inspector's report:

- Insufficient account was taken of the possible tolerance on the roughness coefficients used in the hydraulic design, and moreover research has shown that the methods used give optimistic results because of their neglect of the interaction effects between the shallower berm flows and the deeper main channel flows in two-stage channels.
- In my view it would be indefensible if the scheme were to proceed without reference to the national research programme, the very purpose of which is to provide better information for designing flood channels.
- Some increase in containment bank levels is recommended. Provided these changes are made, the channel will have

sufficient capacity to accommodate an increase in hydraulic
resistance to 40% above the present design assumption.

- I recommend taking into account recent developments in
  understanding and advise re-assessing the performance in this
  light, so that the risks of surprises post-construction are
  minimised.
- To await a large flood before discovering the true capacity rather
  than establishing this under controlled non-emergency conditions would be extremely short-sighted and involve unnecessary
  risk of malfunction when the need is greatest.

In the event, the big flood came just as the scheme was being
completed and it was still very raw and totally untested, so who
can blame the flood managers for deciding to open up the new
river rather than allowing flooding to occur? They would have
been criticised even more by those whose houses were flooded if
they had kept it shut. It is a great shame that it had not been built
to specification but what responsibility the original design consultants had in this respect would, I thought, probably emerge when
the case the Environment Agency brought against its consultants,
Lewin, Fryer & Partners, was heard. This project may well sound
a caution against the Environment Agency requiring flood relief
schemes to give quite such priority to enhancing the environment,
which the Jubilee River has certainly done, as anyone who has
walked along its banks will testify. The hydraulics of the project
must remain sound, whatever the environmental requirements. In
the event, the legal case was settled out of court, with the
consultants bearing much of the cost of remedial works, which
have brought the scheme up to 90% of its design capacity, meaning
that a fifty-year flood can be carried rather than a sixty-year one.
'Nobody is to blame. It was designed in the early 1990s using the
best modelling they had', said a spokesperson for the Environment
Agency. There had been no legal requirement for any of the main
parties to take my advice, of course, even though I had told them
that the scheme had been under-designed because the existing
design methods were defective. This was an example of the
conventional methods behind a long way behind research findings.

CHAPTER 16

In retrospect

MEMOIRS ARE, BY DEFINITION, about what you remember, rather than anything you might have forgotten for one reason or another. So it is just the most memorable happenings on the way through life, and sometimes it is the times of danger that stick in one's mind. Also these are professional memoirs, so say little about my private life except when they became closely linked. My childhood contemporaries could not have had the slightest idea about what the future held for us: there have been such profound changes in our lifetime. We did not imagine that air travel would become commonplace; after all they were the days of Alcock and Brown, and Amy Johnson! Yet we saw the days of supersonic flight – and on one of my many trips to Hong Kong, I filled in a competition form whilst waiting at Heathrow, and received a phone call when I arrived saying that I was one of the winners, the prize being four flights by Concorde, which Margaret and the three grandsons took (the fourth was born later), the trips being to see Santa Claus, flying supersonic over the North Sea! We could never have imagined that we would fly round the world, which Margaret has done once and I have done twice, or that crossing the Atlantic would become so routine. Also those far-sighted ideas of Arthur C Clarke are now commonplace with satellites providing the link for phoning overseas, for the internet and for television programmes. Television itself has developed from the old crystal radios our parents could listen to through headsets. Instead of car ownership being limited to just the richest in society, there are so many cars on the road that we suffer from serious congestion.

Perhaps engineering has changed relatively little apart from the scale of many constructions: bigger, higher dams; longer bridges and tunnels. Thanks to my parents' keenness on me getting the best education possible available to a working-class child and the support from Bootle's education authority, I was able to start my own career in engineering. I have no regrets at all about spending a couple of years helping design aeroplanes, as it was an unforgettable experience. Then I was very lucky to come to Hydraulics Research in Wallingford at just the ideal time, when it was expanding and getting towards a peak in experimental hydraulics. I was fortunate to be able to use all that experience as a consultant in later years, mostly with Binnie & Partners, when that firm was also at its peak. That brought worldwide travel to some exciting places which I have described. So civil engineering has been a wonderful career for me, and I can strongly recommend it to youngsters with a technical bent. Unfortunately, there are unlikely to be any engineers in the generation of our grandchildren, but our son has also enjoyed working in this field, and is now one of the government-approved panel of reservoir inspectors.

The latter half of the twentieth century also saw many other technical changes, for example progressing from slide rules to electronic calculators and computers, with desktop machines now being a must in most households – though not primarily now for computations. Their role has widened to include the video games on which many children waste their time, and improved information sources through the internet and better communication via emails. Even they are becoming old hat with text messaging on mobile phones, the advent of which seems to have made the younger generations feel that they have to talk to someone or other much of the time! I feel that not all these developments have been of great benefit to mankind. My father, just before he died, commented that he was glad to have been alive when men first went to the moon (though that was just before the first moon landings). Since then we have learned so much more about the planets and their satellites, but again is there much real benefit to mankind? The quest for knowledge is a worthy cause, but it has been at enormous expense and there are so many people trying to scratch a living from this planet who could have benefited from some of these funds and resources being diverted their way. Perhaps too the second half of the century was the best time for international travel. I certainly have seen far more of the world than I ever imagined as a small boy in Bootle and there have not been many cases of feeling at any risk from the peoples overseas whom we met. Even Iraq, Algeria and Iran were OK though less welcoming than most other countries. However, these are basically no-go areas nowadays, and are likely to remain so whilst this so-called war on terror continues. Islamic states have increasingly become hostile to Western cultures and to those that do not respect theirs. When our government chose to go to war with Iraq on a trumped-up dossier about weapons of mass destruction, I reckoned it would take a whole generation before we could possibly be friends again – but I am not now so optimistic! It is not just the government that has lost respect, the good name of the country itself for justice and fair play has been lost, with our leaders' support for the illegal imprisonment at Guantanemo Bay and what the Americans euphemistically call 'rendition' of prisoners to foreign countries where torture may be meted out to these poor misguided people. Strangely, this feeling has been very well expressed in a recent article by a Labour politician:

> ... I have reached the stage where I am profoundly ashamed of the Government. Blair's craven support for the extremism of US neo-conservative foreign policy has exacerbated the danger of terrorism and the instability and suffering of the Middle East. He has dishonoured the UK, undermined the UN and international law and helped to make the world a more dangerous place. The erosion of the rule of law and civil liberties has weakened our democracy and increased Muslim alienation. (Clare Short, *The Independent*, 14/9/06)

These are strong words but this does not excuse those extreme members of the Muslim faith for the horrors they have perpetrated, which are now worse in Iraq than anywhere else, with the level of violence between the Shia and Sunni. Nor does it condone the violence used by Saddam Hussein to keep the population subdued before the invasion. Our politicians have a very difficult task ahead of them to restore the balance – and to realise finally that the days of empire are over. We cannot impose our way of life on others by military means.

As we have gone through life, we have increasingly come to regard religion as a side issue, not really for us. We have no objection to people having beliefs that comfort them, but strongly object to those who would seek to impose their own beliefs on others. There have been so many examples through the ages of the cruelty involved in trying to do just that, from the Inquisition and the Crusades to the present day Muslim extremists. We manage very well without religion, although we clearly hold dear the ethical standards in our society which no doubt stem from a Christian upbringing. However, most of those moral beliefs are held also by other religions. Some religions seem to breed sectarianism however, witness Northern Ireland and present day Iraq, and there are so many examples of religious wars, for example between Muslims and Hindus at the time of the partition of India. The religious balance sheet would surely show at least as many disadvantages as benefits, so there would seem to be merit in making the education system entirely secular. It is unfortunate that during our lifetime there has been so little reconciliation between those with different beliefs.

That is the negative side, but the positive side from my personal point of view has been to take satisfaction from one's own achievements. I have been a Fellow of the Institution of Civil Engineers since 1972, which is the highest grade in that Institution, and I am pleased that my son, John, is now also an FICE. Another fellowship I value very much is FCGI, Fellow of the City and Guilds of London Institute, which was the special award from my old London college. It was awarded to those ex-students and staff who had made particularly significant contributions, and each year there is a lunch in the Rector's House in South Kensington that I am pleased to be able to attend. However, it is no longer confined to the engineering and sciences, but has been expanded to include the vocations covered by City and Guilds qualifications, and moreover engineering excellence alone is not an adequate qualification for Fellowship, so its character had changed somewhat. However, it seems that the annual lunches still attract mostly us technologists. Of course I have mainly worked at the interface between science and engineering, and therein lies a tale.

Geoffrey Binnie was very highly respected in the profession and it was a privilege to work with him at times. He was appointed a Fellow of the Royal Society when that august body decided to take on board some engineers to counterbalance their otherwise very academic membership, and he was also one of the first Fellows of the Academy of Engineering when it was set up. He thought I should also be an FEng, so put my name forward. The answer that came back seemed to be, 'This man has published a lot of research papers. He would be more appropriate to a science body.' So his next move was to propose me for Fellowship of the Royal Society but the answer came back, 'This man is more of an engineer than a scientist. FEng might be appropriate.' So there I was between the two, but I believe I achieved more of practical use by providing that link than possibly any other way. Later on, my good friend from the early days, Eddie Naylor, who had had a brilliant career, retiring when he was City Engineer of Manchester and who was himself an FEng, told me he planned to submit my name to the Academy. However, by that time I was around retirement age and the response was that they really wanted to concentrate now on getting in some younger Fellows, a good objective of course. So was I a scientist or an engineer? I like to think that I was both in a modest sort of way.

Global warming is a very live issue at the present, with much correspondence in the press and in the technical journals. It has also become a hot political issue, with party politicians vying with each other to put forward 'green' agenda. But is this doing much more than generating hot air when there are very practical problems to turn our attention to with the rising sea levels and prospects of changing weather patterns? Engineers have an important role to play in this and there is some evidence that action is indeed being taken quietly, away from the headlines, such as the building of refuge platforms above flood level in Bangladesh, and thought being given to raising the flood defences around the south-east shoreline. Also at long last the tidal barrier for Venice is under construction. Many engineers share my view that effort should be concentrated on these practical steps rather than on efforts to curb the emission of greenhouse gases. It seems clear to many that there is no chance whatever of reversing global warming, and recent evidence is that it is happening rather more rapidly than the atmospheric scientists predicted ten years ago. This country has some of the best such people in the world but who actually heeds their findings, such as that even with full achievement of the Kyoto objectives the best to hope for is about a 5% slowing of the process, in other words we will have 105 years rather than 100 years before a given rise in sea levels occurs.

There is much to be said for economy in fuel consumption and the use of renewable sources of power but, with the emphasis on private financing, the Severn Barrage is unlikely to be constructed, and yet there will undoubtedly be hidden subsidies for the next generation of nuclear stations. Carbon dioxide is perhaps the largest anthropogenic source of greenhouse gas but it is easy to forget that 10,000 years ago, as the last ice age came to an end, the sea level was low enough for there to be a land connection between the British Isles and Europe. Global warming was taking place long before the Industrial Revolution, long before man began contributing to it. The atmosphere contains so much carbon dioxide already that additional amounts have perhaps only a fifth the impact they would have if it were a new gas. There are newer man-made gases that have a much worse effect, particle for particle, for example the CFCs, used until recently in fridges, are 14,000 times more potent than CO_2 according to recent research, hence they are now banned and must be removed very carefully from abandoned fridges. There are many poor island states that need help from the wealthy countries to increase their protection or, in the most severe cases, cover their evacuation. There are many challenges for the next generation of engineers to address.

Through membership of IAHR I have made many friends from overseas, several of whom also became Honorary Life Members, including, for example, Paul Novak whom I first met when he came to HR Wallingford, on his arrival in the UK as a refugee from Prague when the Russians invaded that country, in effect to suppress democracy. He and his family had left almost at a moment's notice with what belongings they could get in their car, leaving behind his job as director of the laboratory there and their house and its contents, to be unable to return for many years. He obtained the Chair in Hydraulics at Newcastle, and we have valued their friendship through the years. Many of our IAHR friends we used to meet only every other year, at the congresses, of course. I have also felt privileged to have such good young engineers working in my teams, at Hydraulics Research and then at Binnies – but even that younger generation are now at or beyond retirement age!

So what sort of legacy is there from a lifetime in engineering? The architect Wren has inscribed on his tomb in St Paul's

Cathedral 'If you wish to see my memorial, look around you', but, although I have been involved in very many projects, I have just had a role in their hydraulic design; they are not my work really, they have been team projects. Was it Einstein who said that what he had achieved was by standing on the shoulders of giants? I have done plenty of that but the scientific papers I have written are being outdated. Even the Ackers and White theory of sediment transport is perhaps no longer the best available. However, in making scientific advances more readily available to the practice of engineering, I feel something useful was achieved. What is more, it gave me the opportunity to see so much of this amazing world of ours. What more can one hope for in this life? It has been a rewarding and exciting career, and when you couple that with a wonderful family and happy home, any early hardships are easily forgotten.

My message to any young person reading this is that engineering is a good profession to follow. There are many challenges ahead too.

Bibliography

This list of publications shows where details of many of the studies described may be found, if required.

A theoretical consideration of side weirs as storm overflows, Proc. Inst. Civil Engnrs, London, vol. 6, pp 250–269, 1957.

Resistance of fluids flowing in channels and pipes, Hydraulics Research Paper no 1, HMSO, London, 1958.

Charts for the hydraulic design of channels and pipes, Hydraulics Research Paper no 2, HMSO, London, 1958 (2nd ed. 1963; 3rd ed. 1969).

An investigation into head losses at sewer manholes, Civil Engineering, London, vol 54, pp 882–884 and 1033–1036, 1959.

The calibration of a gauging weir on the River Wandle with the aid of model experiments (with B A Say), Journal Institution Municipal Engineers, London, vol 86 (2), pp 48–56, 1959.

The vortex drop (with E S Crump), Proc. Inst. Civil Engrs, London, vol 16, pp 433–442, 1960.

Comprehensive formulae for critical depth flumes, Water and Water Engineering, London, vol 65, pp 296–306, 1961.

The hydraulic resistance of drainage conduits, Proc. Inst. Civil Engrs, London, vol 19, pp 307–336, 1961.

Tables for the hydraulic design of channels and pipes, Hydraulics Research Paper no 4, HMSO, London, 1963 (2nd ed. 1969).

Critical depth flumes for flow measurement in open channels, Hydraulics Research Paper no 5, HMSO, London, 1963.

Experiments on small streams in alluvium, Proc. American Soc. of Civil Engineers, J. of Hydraulics, vol 90, HY4, pp 1–37, 1964.

Effects of use on the hydraulic resistance of drainage conduits, (with M J Crickmore and D W Holmes), Proc. Inst. Civil Engrs, London, vol 28, pp 339–360, 1964.

Attenuation of flood waves in part-full pipes (with A J M Harrison), Proc. Inst. Civil Engrs, London, vol 28, pp 361–382, 1964.

Estimating the capacity of sewers and storm drains, Municipal Engineering, London, vol 142, pp 170 & 175, 1965.

A study of the effects of flood hydrographs on training works in the meandering Kaduna River, Proc. XI Congress of Int. Ass. for Hydraulic Research, Leningrad, Russia, 1965, paper 3.43.

An investigation of problems associated with high velocity flow in a Jamaican drainage system (with F J T Kestner), Proc. XI Congress of Int. Ass. for Hydraulic Research, Leningrad, Russia, 1965, paper 1.34.

Storm overflow performance studies using crude sewage (with A J Brewer et al.), Proc. symposium on Storm Overflows, Inst. Civil Engrs, London, May, 1967.

Laboratory studies of storm overflows with unsteady flow (with A J M Harrison and A J Brewer), Proc. Symposium on Storm Overflows, Inst. Civil Engrs, London, May, 1967.

The hydraulic design of storm overflows incorporating storage (with A J M Harrison and A J Brewer), Proc. Institution of Municipal Engineers, London, vol 95 (1), pp 31–37, 1968.

Modelling of heated water discharges, Chapter 6, Symposium on Engineering Aspects of Thermal Pollution, Nashville, USA, Vanderbilt Press, Aug 1968.

The geometry of small meandering streams (with F G Charlton), Proc. Inst. Civil Engrs, London, supplement xv, paper 7328S, 1970.

The slope and resistance of small meandering streams (with F G Charlton), Proc. Inst. Civil Engrs, London, supplement xv, paper 7362S, 1970.

Meander geometry arising from varying flows (with F G Charlton), Journal of Hydrology, vol XI, no 3, pp 230–252, Sept 1970.

Dimensional analysis of alluvial channels with special reference to meander length, Journal of Hydraulics Research, vol 8, no 3, pp 283–316, 1970.

Flow measurement by weirs and flumes, paper no 3, International Conference on Modern Developments in Flow Measurement, Harwell, UK, Sept 1971.

The applicability of hydraulic models to pollution studies (with L J Jaffrey), paper 16, Symposium on Mathematical Modelling of Estuarine Pollution, Stevenage, UK, 1972.

River regime: research and application, Journal Inst. Water Engineers, London, vol 26, no 5, pp 257–281, July 1972.

Sediment transport: new approach and analysis (with W R White), Proc. American Soc. of Civil Engineers, J. of Hydraulics, vol 99, HY11, pp 2041–3060, paper 10167, Nov 1973.

Similarity criteria for mobile bed models, Proc. XV Congress of Int. Ass. for Hydraulic Research, Istanbul, Turkey, vol 5, pp 61–64, 1973.

Extreme levels arising from meteorological surges (with T D Ruxton), Proc. 14th Coastal Eng. Conference, Copenhagen, Denmark, vol 1 pp 69–86, June 1974.

Typhoon waves at the High Island Dam, Hong Kong (with J A T Aspden and J L Boyd), Proc. International Symposium on Ocean Wave Measurement and Analysis, New Orleans, USA, voll, pp 523–542, Sept. 1974.

Design and operation of air-regulated siphons for reservoir and head-water control (with A R Thomas), paper A1, Symposium on the Design and Operation of Siphons and Siphon Spillways. British Hydromechanics Res. Ass, London, May 1975.

Wash water storage scheme: engineering implications of the hydraulic model studies; Conference Inst. Civil Engrs, London, Nov 1976.

Field tests of rip-rap slope protection in a shallow coastal area (with R M Young), Proc. XVII Congress of International Ass. for Hydraulic Research, Baden-Baden, Germany, subject Ca,1977.

A mathematical model of the closure problem and permanent operation for tidal power studies (with D J Kluth), Proc. XVII Congress of International Ass. for Hydraulic Research, Baden-Baden, Germany, subject Ca, 1977.

Urban drainage: the effects of sediment on performance and design criteria, International Conference on Urban Storm Drainage, University of Southampton (Pentech Press), March 1978.

River regime and sediment transport, Keynote lecture, International Conference on Water Resources Engineering, Asian Inst. Tech., Bangkok, Thailand, Jan 1978.

Tidal power projects – Australia, Colston Symposium on Tidal Power and Estuary Management, University of Bristol, UK, April, 1978.

Weirs and flumes for flow measurement (with W R White et al) Wiley-Interscience, 327 pages, J Wiley, Chichester, UK, 1978.

Use of sediment transport concepts in stable channel design, Proc. International Workshop in Alluvial River Problems, University of Roorkee, India, March 1980.

Bed material transport: a theory for total load and its verification (with W R White), paper B10, Proc. International Symposium on River Sedimentation, Chinese Soc. of Hydraulic Eng., Beijing, China, March 1980.

Meandering channels and the influence of bed material, International Workshop on Engineering Problems in the Management of gravel bed rivers, Univ. of East Anglia, held at Newtown, UK, (Wiley), June 1980.

Dispersion of cooling water from a coastal LNG plant (with J D Pitt et al), 16th International Conference on Coastal Eng., Sydney, Australia, 1980.

The laminar flow of sewage sludge through pipelines (with M C Allen), The Public Health Engineer, Journal of Inst. of Public Health, London, vol 8, no 3, July 1980.

Barrage operation, flood evacuation, surge tide and closure dynamics, paper no 8, Conference on the Severn Barrage, Inst. Civil Engrs, London, Oct 1981.

Hydraulics Research and Engineering Practice: is there a communication gap, Keynote address, Conf. on Hydraulics in Civil Engineering, Sydney, Australia, Oct 1981.

Novel angled-entry stilling basin (with R W O'Garra), paper BI, International Conference on Hydraulic Modelling of Engineering Structures, Coventry, Sept. 1982 (British Hydromechanics Res. Ass.).

A mathematical model of a floating boom (with D C Keiller) paper J1 International Conference on Hydraulic Modelling of Engineering Structures, Coventry, Sept. 1982 (British Hydromechanics Res. Ass.).

Prototype tests on rip-rap under random wave attack (with J D Pitt), 18th International Conference on Coastal Eng., Capetown, South Africa, Nov 1982.

General method for critical point on spillways (with I R Wood and
  J Loveless), Proc. American Society of Civil Engineers, vol 109,
  HY2, Feb 1983, pp 308–312.
Field scale studies of rip-rap (with J D Pitt), paper 7, International
  Conference on Breakwaters – design and construction, Inst.
  Civil Engrs, London 1 April 1983.
The turbulent and transitional flow of sewage sludge through
  pipelines (with M C Allen), The Public Health Engineer,
  Journal of Inst. of Public Health, London, vol 11, no 3, July
  1983.
Sediment transport problems in irrigation systems, Developments
  in Hydraulic Engineering, vol 1, editor P Novak, Applied
  Science Publishers Ltd, 1983.
Hydraulics Research: Communication and consequences, Unwin
  Lecture, March 1984, Proc. Inst. Civil Engrs, London, part I, vol
  76, Nav 1984, pp 1053–1068.
Segment lined tunnels: field scale tests to determine hydraulic
  roughness (with J D Pitt), 3rd International Conference on
  Urban Storm Drainage, Chalmers University, Gotenburg,
  Sweden, June 1984.
Sediment transport in sewers and the design implications, Confer-
  ence on the Planning, Construction, Maintenance and Oper-
  ation of Sewerage Systems, Reading, UK, Sept. 1984.
Segment lined tunnels: field investigation of roughness, Conference
  on the Planning Construction, Maintenance and Operation of
  Sewerage Systems, Reading, UK, Sept. 1984.
Modelling sediment in rivers, Int. Symposium on Advances in
  Water Engineering, University of Birmingham, July 1985.
Self-aerated flow down a chute spillway (with S J Priestley), 2nd
  International Conference on Floods and Flood Control, Eng-
  land, Sept. 1985, paper A1, BHRA, Cranfield, UK.
Reservoir sedimentation and influence of flushing, Conf. on
  Sediment Transport in Gravel Bed Rivers, John Wiley, 1987, pp
  845–868.
Dimensional analysis, dynamic similarity, process functions, em-
  pirical equations and experience – how useful are they? NATO
  Workshop on movable bed physical models, De Voorst, The
  Netherlands, August 1987.

Scale models: examples of how, why and when – with some ifs and buts; Special Lecture, Proc. XXII Congress of IAHR, Lausanne, Aug–Sept, 1987, Tech Session B, pp 1–16.

Modelling alluvial systems: a review of methods and case studies: Invited Lecture, Gruppo Nazionale Idraulica, Transporto Solido ed Evoluzione Morfologica nel corsa d'acqua, Universita di Trento, Italy, June 1988.

Alluvial channel hydraulics, Journal of Hydrology, vol 100, 1988, Elsevier Science Pub., pp 177–204.

Sediment aspects of drainage and outfall design, Invited Lecture, Proc. International Symposium on Environmental Hydraulics, Hong Kong, Dec 1991; Balkema, Rotterdam, 1991.

Two-stage channels: flow coherence and interactions, Proc. 3rd Int. Conference on Floods and Flood Management, Nov 1992, Florence, Italy, Kluwer Academic Pub., The Netherlands, Fluid Mechanics and its Applications, vol 15, 1992, pp 477–490.

Canal and river regime in theory and practice: 1929–1992; Gerald Lacey Memorial Lecture, Institution of Civil Engineers, London, Water, Maritime and Energy, vol 96, Sept. 1992, pp 167–178.

Hydraulic Design of two-stage channels, Proc. Institution of Civil Engineers, London, Water, Maritime and Energy, vol 96, Dec. 1992, pp 247–257.

Flow formulae for straight two-stage channels, Journal of Hydraulics Research, vol 31, 1993, no 4, pp 509–531.

Sediment transport in open channels: Ackers and White update, Proc. Institution of Civil Engineers, London, Water, Maritime and Energy, vol 101, Dec 1993, pp 247–24.

Index

Washington State University 23

Places and projects